MEI Structured Mathematics

Mathematics is not only a beautiful and exciting subject in its own right but also one that underpins many other branches of learning. It is consequently fundamental to the success of a modern economy.

MEI Structured Mathematics is designed to increase substantially the number of people taking the subject post-GCSE, by making it accessible, interesting and relevant to a wide range of students.

It is a credit accumulation scheme based on 45 hour units which may be taken individually or aggregated to give Advanced Subsidiary (AS) and Advanced GCE (A Level) qualifications in Mathematics and Further Mathematics. The units may also be used to obtain credit towards other types of qualification.

The course is examined by OCR (previously the Oxford and Cambridge Schools Examination Board) with examinations held in January and June each year.

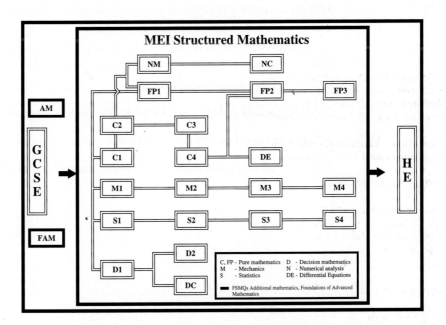

This is one of the series of books written to support the course. Its position within the whole scheme can be seen in the diagram above.

Mathematics in Education and Industry (MEI) is an independent curriculum development body which aims to promote links between education and industry in mathematics. MEI produce relevant examination specifications at GCSE, AS and A Level (including Further Mathematics) and for Free Standing Mathematics Qualifications (FSMQs); these are examined by OCR.

In partnership with Hodder Murray, MEI are responsible for three major series of textbooks: Formula One Maths for Key Stage 3, Hodder Mathematics for GCSE and the MEI Structured Mathematics series, including this book, for AS and A Level.

As well as textbooks, MEI take a leading role in the development of on-line resources to support mathematics. The books in this series are complemented by a major MEI website providing full solutions to the exercises, extra questions including on-line multiple choice tests, interactive demonstrations of the mathematics, schemes of work, and much more.

In recent years MEI have worked hard to promote Further Mathematics and, in conjunction with the DfES, they are now establishing the national network of Further Mathematics Centres.

MEI are committed to supporting the professional development of teachers. In addition to a programme of Continual Professional Development, MEI, in partnership with several universities, co-ordinate the Teaching Advanced Mathematics programme, a course designed to give teachers the skills and confidence to teach A Level mathematics successfully.

Much of the work of MEI is supported by the Gatsby Charitable Foundation.

MEI is a registered charity and a charitable company.

MEI's website and email addresses are www.mei.org.uk and office@mei.org.uk.

MEI STRUCTURED MATHEMATICS

THIRD EDITION

Decision Mathematics 2 and C

Keith Parramore
Joan Stephens
Chris Compton

Series Editor: Roger Porkess

Hodder Murray
A MEMBER OF THE HODDER HEADLINE GROUP

Acknowledgements

We are grateful to the following companies, institutions and individuals who have given permission to reproduce photographs in this book. Every effort has been made to trace and acknowledge ownership of copyright. The publishers will be glad to make suitable arrangements with any copyright holder it has not been possible to contact.

Bettmann/CORBIS (page 111; CORBIS (ih000935) (page 112); Life File (page 126); Bill Varie/CORBIS (page 126)

OCR, AQA and Edexcel accept no responsibility whatsoever for the accuracy or method of working in the answers given.

Orders: please contact Bookpoint Ltd, 78 Milton Park, Abingdon, Oxon OX14 4TD. Telephone: (44) 01235 827720, Fax: (44) 01235 400454. Lines are open from 9 am to 5 pm, Monday to Saturday, with a 24-hour message-answering service. You can also order through our website at *www.hoddereducation.co.uk*.

British Library Cataloguing in Publication Data
A catalogue record for this title is available from the The British Library

ISBN-13: 978 0 340 889992

First Edition Published 1992
Second Edition Published 2001
Third Edition Published 2004
Impression number   10 9 8 7 6 5 4 3 2
Year   2010 2009 2008 2007

Hodder Headline's policy is to use papers that are natural, renewable and recyclable products and made from wood grown in sustainable forests. The logging and manufacturing processes are expected to conform to the environmental regulations of the country of origin.

Typeset by Pantek Arts Ltd, Maidstone, Kent.
Printed in India for Hodder Education a member of the Hodder Headline Group, 338 Euston Road, London NW1 3BH.

Key to symbols in this book

? This symbol means that you may want to discuss a point with your teacher. If you are working on your own there are answers in the back of the book. It is important, however, that you have a go at answering the questions before looking up the answers if you are to understand the mathematics fully.

⚠ This is a warning sign. It is used where a common mistake, misunderstanding or tricky point is being described.

▢ This is the ICT icon. It indicates where you should use a graphic calculator or a computer.

e This symbol and a dotted line down the right-hand side of the page indicates material which is beyond the criteria for the unit but which is included for completeness.

☆☆ Harder questions are indicated with stars. Many of these go beyond the usual examination standard.

Introduction

This is the second of two volumes to support the Decision Mathematics modules in MEI Structured Mathematics. The two volumes together cover the basic work in this branch of mathematics and so are suitable for use with any decision mathematics course at this level.

Chapters 1 to 4 of this volume cover the material required for the Decision Mathematics 2 module. Chapters 5 to 8 cover the material for the Decision Mathematics Computation module.

The linear programming in Chapter 1 covers theoretical aspects. That in Chapter 6 is focused more on applications, particularly on computing applications relating to the network problems in Chapter 5.

The topics are split similarly. Logic in Chapter 4 provides a theoretical underpinning for computing, whilst Chapter 7 (Recurrence relations) uses a computing approach to back up the theory. Decision analysis in Chapter 3 does not require computing, whilst the simulation in Chapter 8 does.

In both sections the underlying theme is *modelling* (see the diagram on the next page). Decision mathematics is, broadly, the application of mathematical modelling to solve real world problems, often arising from managing commercial and industrial concerns. To use the power of mathematics to solve problems you first need to capture the essence of the real world problem in mathematical form. This move from the real world into the world of mathematics is known as mathematical modelling. It requires simplifying assumptions – so that the mathematical problem which is extracted is tractable.

There are many specific models covered throughout the book. It is to be hoped that, as well as learning about these models, students will improve their modelling skills. Even when the problem and model are both well defined there are often questions to be asked about assumptions, range of applicability, practicality, etc. Examination questions often have sections exploring such aspects.

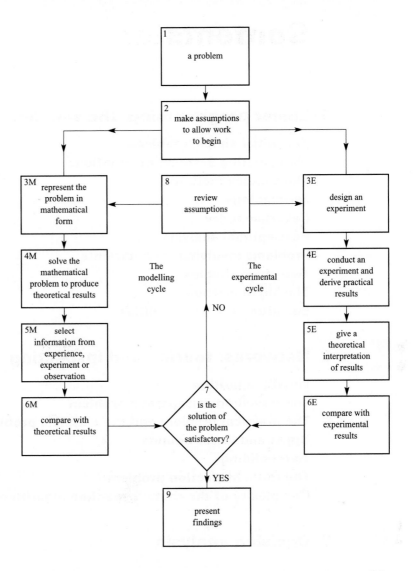

The examination for Decision Mathematics Computation requires candidates to have access to a computer, with a spreadsheet and a linear programming package, allowing realistic problems to be solved.

This is the third edition of this volume and it contains many additional examination questions. We would like to thank Ian Bloomfield for his helpful comments on the material. We would also like to thank the various examination boards who have given permission for their past questions to be included in the exercises.

Keith Parramore and Joan Stephens

Contents

1 Linear programming: the simplex method

Simplex munditiis (so simple, so neat)

Horace

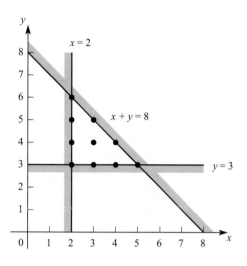

Figure 1.1

Figure 1.1 illustrates the inequalities $x \geqslant 2$, $y \geqslant 3$ and $x + y \leqslant 8$.

Suppose that x and y must be integers.

❓

1 How many points are there in the feasible region?

2 Which of the feasible points gives a maximum value of $2x + 3y$?

3 Now think about this problem for integers x, y and z. What is the greatest value of $x + 2y + 3z$ given that $x \geqslant 1$, $y \geqslant 2$, $z \geqslant 3$ and $2x + 3y + z \leqslant 25$?

4 Why is it difficult to use the method illustrated in figure 1.1?

5 How did you do it?

Now look at this problem about drinks production.

The day's output of two drinks from a production unit is to be planned. Each 5 litres of the energy drink requires 1.25 litres of syrup, 2 units of vitamin supplement and 30 cm³ of concentrated flavouring. Each 5 litres of the refresher drink also requires 1.25 litres of syrup, but only 1 unit of vitamin supplement and 20 cm³ of flavouring. There are 250 litres of syrup available, 300 units of vitamin supplement and 4.8 litres of concentrated flavouring.

The energy drink sells at £1 per litre. The refresher drink sells at 80 pence per litre. How much should be produced of each drink to maximise the day's income?

You could answer this with a graph because there are only two variables. (This is done in Chapter 5 of *Decision Mathematics 1*.) You will now solve it using an algebraic approach, so that you can then solve more complex problems with more than two variables.

In *Decision Mathematics 1* this was formulated as an LP problem:

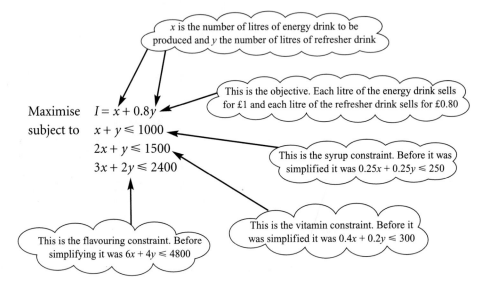

Maximise $I = x + 0.8y$
subject to $x + y \leqslant 1000$
             $2x + y \leqslant 1500$
             $3x + 2y \leqslant 2400$

x is the number of litres of energy drink to be produced and y the number of litres of refresher drink

This is the objective. Each litre of the energy drink sells for £1 and each litre of the refresher drink sells for £0.80

This is the syrup constraint. Before it was simplified it was $0.25x + 0.25y \leqslant 250$

This is the vitamin constraint. Before it was simplified it was $0.4x + 0.2y \leqslant 300$

This is the flavouring constraint. Before simplifying it was $6x + 4y \leqslant 4800$

The graphical solution is shown in figure 1.2.

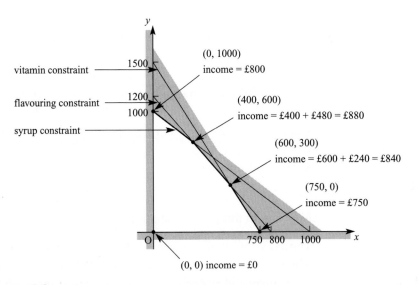

Figure 1.2

Since the best solution is at a vertex of the feasible region, the graph shows that the best solution is to produce 400 litres of the energy drink and 600 litres of the refresher drink, giving an income of £880.

The initial simplex tableau

In *Decision Mathematics 1* the constraint inequalities were written as equalities by using slack variables. The drink production problem was rewritten as:

> The objective was reorganised so that it looked like the constraints, with variables collected on the left-hand side

Maximise I where $\quad I - x - 0.8y = 0$

subject to
$$x + y + s_1 = 1000$$
$$2x + y + s_2 = 1500$$
$$3x + 2y + s_3 = 2400$$

> s_1, s_2 and s_3 are called slack variables. Their values depend on how much of each resource, syrup, vitamin supplement and flavouring, remains unused

⚠ The unsimplified syrup constraint was $0.25x + 0.25y \leqslant 250$. As this was scaled up by a factor of four to make it easier to handle, the slack variable s_1 actually represents four times the number of litres of unused syrup. Similarly s_2 is five times the number of units of unused vitamin supplement and s_3 is half the number of cubic centimetres of unused flavouring.

The equations are usually set out in tabular form, known as a *simplex tableau*, as shown below.

I	x	y	s_1	s_2	s_3	RHS
1	−1	−0.8	0	0	0	0
0	1	1	1	0	0	1000
0	2	1	0	1	0	1500
0	3	2	0	0	1	2400

You should make sure that you can identify the lines of the tableau with the objective and the constraints in the original formulation.

I	x	y	s_1	s_2	s_3	RHS
1	−1	−0.8	0	0	0	0
0	1	1	1	0	0	1000
0	2	1	0	1	0	1500
0	3	2	0	0	1	2400

Maximise $\quad I = x + 0.8y$
subject to $\quad x + y \leqslant 1000$
$\quad\quad\quad\quad 2x + y \leqslant 1500$
$\quad\quad\quad\quad 3x + 2y \leqslant 2400$

Manipulating systems of equations

You will be manipulating the simplex tableau in exactly the same way as you manipulate systems of equations.

For instance, the system

$$2x + 4y = 6 \qquad \text{①}$$
$$3x + 8y = 7 \qquad \text{②}$$

is equivalent to the system

$$x + 2y = 3 \quad \text{(scaling equation ① by 0.5)} \qquad \text{③}$$
$$3x + 8y = 7$$

which, in turn, is equivalent to

$$x + 2y = 3$$
$$2y = -2 \quad \text{(subtracting } 3 \times \text{equation ③} \qquad \text{④}$$
$$\text{from equation ②)}$$

and finally, the system is equivalent to

$$x = 5 \quad \text{(subtracting equation ④}$$
$$y = -1 \quad \text{from equation ③ and}$$
$$\text{scaling equation ④ by 0.5)}$$

Equivalent means that at each stage the system of equations (in this case two equations in two unknowns) always has the same solution. In this case $x = 5$ and $y = -1$ always fits. Of course the final system of two equations in two unknowns is rather special, since it is the solution.

Scaling equations in a system or combining them together like this (known as taking linear combinations) always produces equivalent systems.

? What happens if the system of equations consists of two equations in three unknowns?

Try it with the system
$$4x + y + 2z = 11$$
$$3x + y + z = 7.$$

Show that the system
$$2x + y = 3$$
$$x + z = 4 \quad \text{is equivalent.}$$

When the number of unknowns exceeds the number of equations it is usually the case that there are many solutions to the system (unless one equation is just a linear combination of the others). In the example above you can choose any value of x you like, and work out corresponding values of y and z to give a solution. An obvious choice (but not the only one) is to choose $x = 0$, giving $y = 3$ and $z = 4$.

In the drink production problem rewriting the constraints produced three equations in five unknowns:

$$x + y + s_1 = 1000$$
$$2x + y + s_2 = 1500$$
$$3x + 2y + s_3 = 2400$$

There are many solutions to this system. An obvious one is $x = 0$, $y = 0$, $s_1 = 1000$, $s_2 = 1500$ and $s_3 = 2400$. This is not a very useful solution to this problem as it means making no drinks and using none of the resources, but it is a start!

The simplex method

Dantzig developed the simplex method in the 1940s.

He called his method the *simplex* method because the proofs underpinning it depend on the convexity of the feasible region, and the simplest convex shape in n dimensions is called a simplex. It is defined by $n + 1$ equally spaced points.

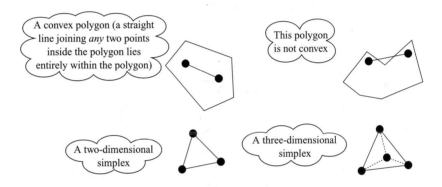

Figure 1.3

Historical note

George Dantzig was born in Oregon, USA, in 1914. As a young man he worked as a statistician for the US government but during the second world war he became head of the Combat Analysis branch of the air force. It was there that he became an expert on planning methods, work that he later developed into the simplex method of optimisation (published in 1947). Since then Dantzig has continued to work with distinction in the developing field of Operational Research.

Dantzig's method is an iterative method. It reorganises the constraint equations at each step so that a new, improved, 'obvious' solution can be seen. It starts with the initial tableau, which has as its obvious solution the point corresponding to the origin of the graph.

I	x	y	s_1	s_2	s_3	RHS
1	−1	−0.8	0	0	0	0
0	1	1	1	0	0	1000
0	2	1	0	1	0	1500
0	3	2	0	0	1	2400

Maximise $I = x + 0.8y$

subject to
$$x + y + s_1 = 1000$$
$$2x + y + s_2 = 1500$$
$$3x + 2y + s_3 = 2400$$

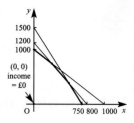

Figure 1.4

The algorithm

An iteration of the algorithm is known as a *pivot*. It has three steps: choosing the pivot column, choosing the pivot element and making the pivot. During all this no variable is ever allowed to take a negative value, that is, entries in the RHS column are always non-negative.

Choosing the pivot column

This involves examining the objective row:

I	x	y	s_1	s_2	s_3	RHS
1	−1	−0.8	0	0	0	0

This is read as $I - x - 0.8y = 0$ or $I = x + 0.8y$. In the current solution x and y both have value 0. Hence I has value 0. Increasing either x or y will increase the value of I.

In general this analysis shows that if there is a negative entry in the objective row, then improvement is possible by increasing the value of the corresponding variable. Where there is a choice it is conventional (but not necessary) to choose the variable corresponding to the most negative entry. The corresponding column is called the pivot column.

In the drinks production problem the pivot column for the first iteration will be the x column (although it could have been the y column).

Choosing the pivot element

Having chosen to increase x you would like to increase it by as much as is possible. To see how much this is, consider the constraint equations:

$$x + y + s_1 \quad = 1000$$
$$2x + y \quad + s_2 \quad = 1500$$
$$3x + 2y \quad + s_3 = 2400$$

Looking at the first equation you can see that if you increase x then s_1 will have to decrease to maintain the balance. You can increase x to 1000, making $s_1 = 0$. A greater increase would make s_1 negative, which is not allowed. The second equation limits you to increase x to no more than 750, as otherwise s_2 would be negative. The third equation imposes a limit of 800. Taken together you can see that the maximum possible value of x is 750, and that increasing x to this value will make s_2 zero.

In terms of the tableau, what you have done involves dividing the RHS of each row by the corresponding entry in the pivot column, provided that entry is positive.

? You can ignore entries which are zero or negative. Can you see why?

The minimum of the ratios so obtained defines the pivot element. In the example the ratios are $\frac{1000}{1} = 1000$, $\frac{1500}{2} = 750$ and $\frac{2400}{3} = 800$, giving the entry in the second constraint row (or the third row of the table) as the pivot element. This is called the *ratio test*.

Making the pivot

You now manipulate the system of equations so that the pivot element becomes 1, and the remaining elements of the pivot column all become 0. The manipulations which you need are shown below.

Divide through the pivot row by the pivot element.

I	x	y	s_1	s_2	s_3	RHS
1	−1	−0.8	0	0	0	0
0	1	1	1	0	0	1000
0	①	0.5	0	0.5	0	750
0	3	2	0	0	1	2400

Pivot column points to the x column. Objective row is the first row (RHS 0). Pivot row is the third row (RHS 750).

Note that the pivot element is circled. The second row has already been divided through by 2.

Add the pivot row to the objective row to create a 0 in the pivot column.

I	x	y	s_1	s_2	s_3	RHS
1	0	−0.3	0	0.5	0	750
0	1	1	1	0	0	1000
0	①	0.5	0	0.5	0	750
0	3	2	0	0	1	2400

Subtract the pivot row from the first constraint row to create a 0 in the pivot column.

I	x	y	s_1	s_2	s_3	RHS
1	0	−0.3	0	0.5	0	750
0	0	0.5	1	−0.5	0	250
0	①	0.5	0	0.5	0	750
0	3	2	0	0	1	2400

Subtract 3 times the pivot row from the final row to create a 0 in the pivot column.

I	x	y	s_1	s_2	s_3	RHS
1	0	−0.3	0	0.5	0	750
0	0	0.5	1	−0.5	0	250
0	1	0.5	0	0.5	0	750
0	0	0.5	0	−1.5	1	150

As with the initial tableau there is an 'obvious' solution. The x column contains zeros and a single 1. You can put x equal to the RHS value of the row containing that 1, i.e., 750. Independently of that you can put s_1 equal to 250, since its single 1 is in the row which has RHS = 250. Finally you can, without affecting x or s_1, put s_3 equal to 150. Setting both the remaining variables, y and s_2 to zero completes the solution. This gives $I = 750$ and corresponds to (750, 0) in the graphical solution.

The fact that $s_2 = 0$ means that the solution involves using all of the vitamin supplement.

I	x	y	s_1	s_2	s_3	RHS
1	0	−0.3	0	0.5	0	750
0	0	0.5	1	−0.5	0	250
0	1	0.5	0	0.5	0	750
0	0	0.5	0	−1.5	1	150

Maximise $I = 750 + 0.3y - 0.5s_2$

subject to $0.5y + s_1 - 0.5s_2 = 250$

$x + 0.5y + 0.5s_2 = 750$

$0.5y - 1.5s_2 + s_3 = 150$

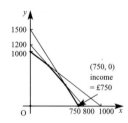

Figure 1.5

Further pivots

The objective row now reads $I - 0.3y + 0.5s_2 = 750$.

The negative y value shows that, whilst I has been increased to 750, further improvement is possible by increasing y.

In general a negative entry in the objective row shows that at least one further pivot is necessary. Its position defines the pivot column, in this case the y column.

The ratio test gives $\frac{250}{0.5} = 500$, $\frac{750}{0.5} = 1500$ and $\frac{150}{0.5} = 300$. The minimum is 300, showing that the pivot element is the final entry in the pivot column.

You can now make the second pivot.

Divide through the pivot row by the pivot element, 0.5 (i.e. multiply through by 2).

I	x	y	s_1	s_2	s_3	RHS
1	0	−0.3	0	0.5	0	750
0	0	0.5	1	−0.5	0	250
0	1	0.5	0	0.5	0	750
0	0	①	0	−3	2	300

Add 0.3 × the pivot row to the objective row to create a 0 in the pivot column.

I	x	y	s_1	s_2	s_3	RHS
1	0	0	0	−0.4	0.6	840
0	0	0.5	1	−0.5	0	250
0	1	0.5	0	0.5	0	750
0	0	①	0	−3	2	300

Subtract 0.5 × the pivot row from the first constraint row to create a 0 in the pivot column.

I	x	y	s_1	s_2	s_3	RHS
1	0	0	0	−0.4	0.6	840
0	0	0	1	1	−1	100
0	1	0.5	0	0.5	0	750
0	0	①	0	−3	2	300

Subtract 0.5 × the pivot row from the second constraint row to create a 0 in the pivot column.

I	x	y	s_1	s_2	s_3	RHS
1	0	0	0	−0.4	0.6	840
0	0	0	1	1	−1	100
0	1	0	0	2	−1	600
0	0	1	0	−3	2	300

You can now interpret the tableau.

I	x	y	s_1	s_2	s_3	RHS
1	0	0	0	−0.4	0.6	840
0	0	0	1	1	−1	100
0	1	0	0	2	−1	600
0	0	1	0	−3	2	300

Maximise $I = 840 + 0.4s_2 - 0.6s_3$

subject to

$$s_1 + s_2 - s_3 = 100$$
$$x + 2s_2 - s_3 = 600$$
$$y - 3s_2 + 2s_3 = 300$$

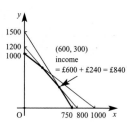

Figure 1.6

At this point, the end of the second iteration, there is a new 'obvious' solution. It is $x = 600$, $y = 300$, $s_1 = 100$, $s_2 = 0$ and $s_3 = 0$ with $I = 840$. This corresponds to (600, 300) in the graphical solution. The fact that $s_2 = 0$ and $s_3 = 0$ means that the solution involves using all of the vitamin supplement and flavouring. 25 litres of syrup remain unused (because s_1 is four times the number of litres of syrup unused, see page 3).

 The original simplification of the constraints makes it more difficult to interpret non-zero slack variables. Users of LP computer packages avoid making simplifications since computers can cope easily with awkward numbers!

 How can you tell that this solution is not the optimal solution?

You can see that the current solution is not optimal because there is a negative entry, −0.4, in the objective row. This implies that you can further increase I by increasing s_2.

So in the next iteration the pivot column is the s_2 column. The ratio test gives $\frac{100}{1}$ and $\frac{600}{2}$ (you do not need to consider 300 divided by −3, since it is negative).

The minimum is 100 so the pivot element is in the second row of the table. You should find that the pivot gives:

I	x	y	s_1	s_2	s_3	RHS
1	0	0	0.4	0	0.2	880
0	0	0	1	①	−1	100
0	1	0	−2	0	1	400
0	0	1	3	0	−1	600

The tableau is now optimal since there are no more negative entries in the pivot row. The optimal solution is $x = 400$, $y = 600$, $s_1 = 0$, $s_2 = 100$ and $s_3 = 0$, with $I = 880$. The fact that $s_1 = 0$ and $s_3 = 0$ shows that all of the syrup and all of the flavouring is used. There are 20 units of vitamin supplement left over (because s_2 is five times the number of units unused, see page 3).

I	x	y	s_1	s_2	s_3	RHS
1	0	0	0.4	0	0.2	880
0	0	0	1	1	-1	100
0	1	0	-2	0	1	400
0	0	1	3	0	-1	600

Maximise $I = 880 - 0.4s_1 - 0.2s_3$

subject to
$$s_1 + s_2 - s_3 = 100$$
$$x - 2s_1 + s_3 = 400$$
$$y + 3s_1 - s_3 = 600$$

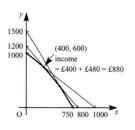

Figure 1.7

 What would have happened if, at the first iteration, you had chosen to increase y rather than x, that is, if you had chosen the y column to be the pivot column?

The solutions given by the pivots are all vertices of the feasible region. The sequence of solutions is called a *trajectory*. In this example the trajectory was $(0, 0)$ $\rightarrow (750, 0) \rightarrow (600, 300) \rightarrow (400, 600)$. Had you chosen to increase y first the trajectory would have been $(0, 0) \rightarrow (0, 1000) \rightarrow (400, 600)$. This would have involved only two iterations, and so less work, but there is no way of knowing in advance which is the best pivot column to choose. In large problems the choice does not make a significant difference to the number of iterations that are needed.

 Throughout the following exercise either the physical situation or the specification of the question requires that the variables be non-negative. The simplex method assumes that all variables can take only non-negative values, so this is dealt with automatically. If a variable x, for instance, should be required to take negative as well as non-negative values, then it can be replaced by $x_1 - x_2$, where x_1 and x_2 are both non-negative.

EXERCISE 1A

1 Use the simplex algorithm to solve the drinks production problem but start by using the y column as the first pivot column.

Maximise $I = x + 0.8y$
subject to $x + y \leqslant 1000$
$$2x + y \leqslant 1500$$
$$3x + 2y \leqslant 2400$$

2 Use the simplex algorithm to solve the following LP.

Maximise $P = 16x + 24y$
subject to $2x + 3y \leqslant 24$
$$2x + y \leqslant 16$$
$$y \leqslant 6$$
$$x \geqslant 0, y \geqslant 0$$

3 Use the simplex algorithm to solve the following LP.

Maximise $P = 9x + 10y + 6z$
subject to $2x + 3y + 4z \leqslant 3$
$6x + 6y + 2z \leqslant 8$
$x \geqslant 0, y \geqslant 0, z \geqslant 0$

4 Use the simplex algorithm to solve the following LP.

Maximise $P = 3w + 2x$
subject to $w + x + y + z \leqslant 150$
$2w + x + 3y + 4z \leqslant 200$
$w \geqslant 0, x \geqslant 0, y \geqslant 0, z \geqslant 0$

5 Use the simplex algorithm to solve the following LP.

Maximise $P = 3w + 2x$
subject to $w + x + y + z \leqslant 150$
$2w + x + 3y + 4z \leqslant 200$
$w \geqslant x$ (rewrite this as $x - w \leqslant 0$)
$w \geqslant 0, x \geqslant 0, y \geqslant 0, z \geqslant 0$

Terminology

In the drinks production example there were five variables: two *state variables*, x and y, and three *slack variables*, s_1, s_2, and s_3. There were three constraints. At the end of each pivot the tableau had a structure in which the first column and three other columns contained zeros together with a single 1. Forgetting the first column, which is always the same, the variables corresponding to these columns are called *basic variables*. The other variables are called *non-basic*. The 'obvious' solution is obtained by making the non-basic variables zero and by giving each basic variable the RHS value of the row containing its 1.

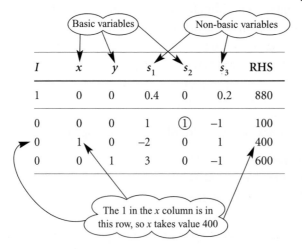

⚠ You should relate these rules for the values taken by basic and non-basic variables to the underlying equations:

$$\text{Maximise } I = 880 - 0.4s_1 - 0.2s_3$$

$$
\begin{aligned}
\text{subject to} \qquad & s_1 + s_2 - s_3 &&= 100 \\
x \quad & -2s_1 \quad + s_3 &&= 400 \\
y + 3s_1 \quad & -s_3 &&= 600
\end{aligned}
$$

A basic variable appears in only one constraint equation, and with coefficient 1. So a solution to the system can be obtained by setting each basic variable to the value of the corresponding entry in the RHS column, and by setting each of the other variables to value zero.

In general there will be a slack variable for each constraint, plus as many state variables as are needed to model the problem. At the end of each iteration the number of basic variables will be equal to the number of constraints and the remainder will be non-basic.

Hypergeometry

You can easily imagine extending the graphical solution method to three variables. The feasible region would be bounded by planes instead of lines so it would be a polyhedron rather than a polygon.

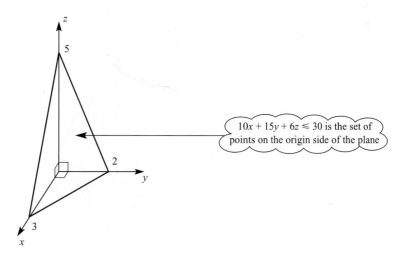

$10x + 15y + 6z \leqslant 30$ is the set of points on the origin side of the plane

Figure 1.8 *The plane 10x + 15y + 6z = 30.*

You can imagine several such planes defining a polyhedral feasible region containing the origin down in its bottom corner, such as is shown in figure 1.9.

You can argue that the best solution must lie at a vertex (exceptionally at more than one vertex) of your feasible region. However, there would now tend to be more vertices to check for a given number of constraints. Furthermore finding a vertex would involve solving three linear equations in three unknowns and you would not actually be able to do it by drawing it!

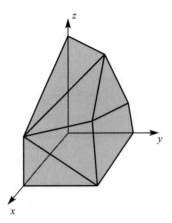

Figure 1.9

For more than three variables it is difficult to picture the situation. However, the algebra that describes the lines and polygons in two dimensions and the planes and polyhedra in three dimensions, behaves exactly the same in any number of dimensions. A realistic linear programming problem might have tens, or even hundreds, of variables. The constraint planes are then called *hyperplanes*.

The simplex algorithm starts at the origin and moves in a trajectory along edges of the n-dimensional feasible region, always in the direction of improving the objective, until it reaches the optimal vertex.

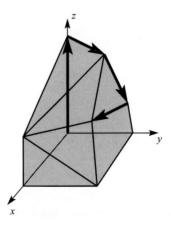

Figure 1.10

Post-optimal analysis

The final tableau contains further information. Here is the final tableau for the drinks production problem.

I	x	y	s_1	s_2	s_3	RHS
1	0	0	0.4	0	0.2	880
0	0	0	1	1	−1	100
0	1	0	−2	0	1	400
0	0	1	3	0	−1	600

You know what each of the figures in the RHS column means. The 880 gives the optimal value of the objective, in this case representing the optimal income in pounds. The 100 shows that there is vitamin supplement left, in this case 20 units, since s_2 is five times the number of units of vitamin supplement left unused. The 400 and 600 are the optimal values of x and y, in this case the number of litres of each drink that should be produced.

The numbers in the objective row are also meaningful. 0.4 and 0.2 give indications of the benefits to be gained by relaxing the tight constraints (in this case by buying in more syrup or more flavouring). The precise interpretation is made more difficult by the earlier simplifications. You will examine this in Chapter 6 if you go on to study Decision Mathematics Computation.

It is also possible to use entries in the final tableau to compute what changes to the original problem would lead to a solution which is different in nature to the current optimal solution. For example, what change in price of the refresher drink would make (400, 600) no longer the optimal production plan. Again this is examined in Chapter 6.

ACTIVITY

What happens to the simplex method when the objective function is parallel to a constraint line, that is, when the solution to the LP is not unique? Investigate this with the following problem.

$$\text{Maximise} \quad P = 8x + 6y$$
$$\text{subject to} \quad 4x + y \leqslant 40$$
$$4x + 3y \leqslant 48$$
$$20x + 31y \leqslant 400$$
$$x \geqslant 0, y \geqslant 0$$

Increase x first when you apply the simplex algorithm. (Note that the objective function is parallel to one of the constraints.)

In the activity above you should have found that in the final tableau two of the non-basic variables had an entry of zero in the objective row. If one or more of the non-basic variables have an entry of zero in the objective row this indicates that another pivot is possible, leading to a new solution in which the value of the objective will not be changed (see Exercise 1A, question 2).

Problems involving ⩾ constraints

The simplex method involves iterations which move from a feasible vertex to an improved feasible vertex. In the examples seen so far the constraints have all been of the ⩽ form, for example, $x + y \leqslant 1000$. This means that the origin has always been a feasible solution and so could serve as a starting point for the process. However, if there is a constraint of the ⩾ form, for example, $x + y \geqslant 800$, then the origin will not be feasible and you will need some other feasible point from which to begin the process.

Suppose that in the drinks production problem the company is required to produce at least 800 litres of drink in total, to satisfy contracted demand. The problem becomes:

$$\text{Maximise} \quad I = x + 0.8y$$
$$\text{subject to} \quad x + y \leqslant 1000$$
$$2x + y \leqslant 1500$$
$$3x + 2y \leqslant 2400$$
$$x + y \geqslant 800$$

Since this particular problem has only two variables you could add the line $x + y = 800$ to your graph and shade the origin side of it to give the feasible region. You would be able to see immediately that this particular constraint does not affect the solution.

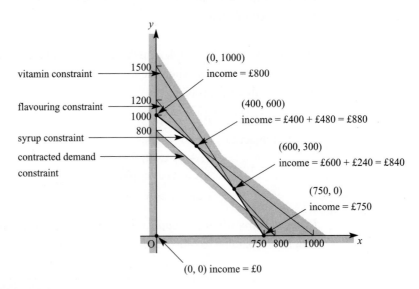

Figure 1.11

With more than two or three variables you would not be able to do this. You need to be able to modify the simplex algorithm so that it can either find an initial feasible solution or do without one in some way. You will examine two ways of doing this:

- two-stage simplex
- the big M method.

Two-stage simplex

In producing a tableau for the extended problem the extra constraint, $x + y \geqslant 800$, creates a problem. To convert it into an equality you need a *surplus* variable s_4, not a slack variable:

$$x + y - s_4 = 800$$

The extra row in the table generates an extra column, the s_4 column.

I	x	y	s_1	s_2	s_3	s_4	RHS
1	−1	−0.8	0	0	0	0	0
0	1	1	1	0	0	0	1000
0	2	1	0	1	0	0	1500
0	3	2	0	0	1	0	2400
0	1	1	0	0	0	−1	800

The extra column is not of the required form, however. It contains −1 and zeros instead of 1 and zeros. Thus you cannot spot an obvious solution since $x = 0$, $y = 0$, $s_1 = 1000$, $s_2 = 1500$, $s_3 = 2400$ and $s_4 = -800$ is not feasible. Variables are not allowed to take negative values.

The solution is to introduce another variable called an *artificial* variable:

$$x + y - s_4 + a_1 = 800$$

The tableau is then as follows.

I	x	y	s_1	s_2	s_3	s_4	a_1	RHS
1	−1	−0.8	0	0	0	0	0	0
0	1	1	1	0	0	0	0	1000
0	2	1	0	1	0	0	0	1500
0	3	2	0	0	1	0	0	2400
0	1	1	0	0	0	−1	1	800

There is now an obvious solution:

$$x = y = s_4 = 0$$
$$s_1 = 1000 \qquad s_2 = 1500$$
$$s_3 = 2400 \qquad a_1 = 800$$
$$I = 0$$

But the solution is, as the name of the new variable suggests, artificial. It does not represent a solution to the original problem, that is, to the problem without the artificial variable. However you can turn this into a simplex tableau for a new problem by adding a new objective, minimise $A = a_1$.

Note

In cases in which there is more than one $\geqslant$ constraint this new objective will be of the form $A = a_1 + a_2 + ... + a_n$.

Until now you have always maximised the objective in LP problems. Now you will also need to minimise this new objective.

You will want to preserve a_1 as a basic variable so you do not want another non-zero entry to appear in the a_1 column. To avoid this you must use $x + y - s_4 + a_1 = 800$ to rephrase this new objective.

$$A = a_1 \qquad = 800 - x - y + s_4$$
$$\text{so} \quad A + x + y - s_4 = 800$$

This gives the following tableau.

A	I	x	y	s_1	s_2	s_3	s_4	a_1	RHS
1	0	1	1	0	0	0	−1	0	800
0	1	−1	−0.8	0	0	0	0	0	0
0	0	1	1	1	0	0	0	0	1000
0	0	2	1	0	1	0	0	0	1500
0	0	3	2	0	0	1	0	0	2400
0	0	1	1	0	0	0	−1	1	800

The first row represents the new objective, which you will use in the first stage. It gives the sum of the artificial variables (in this case just the one).

The second row gives the original objective, which will come into action in the second stage. The remaining rows are the constraints. Iterations will be exactly as before except that in the first stage, in which you are *minimising* the new objective, *positive* values in the objective row indicate that improvement is possible.

Thus in the first iteration you can choose to increase x or y, since both have positive coefficients in the first objective row. Make the arbitrary choice of increasing x.

You find the pivot element by using the ratio test. This compares $\frac{1000}{1} = 1000$, $\frac{1500}{2} = 750$, $\frac{2400}{3} = 800$ and $\frac{800}{1} = 800$. So the pivot element is the 2 in the second constraint row. You are now in a position to make the first pivot.

Divide through the pivot row by 2.

A	I	x	y	s_1	s_2	s_3	s_4	a_1	RHS
1	0	1	1	0	0	0	-1	0	800
0	1	-1	-0.8	0	0	0	0	0	0
0	0	1	1	1	0	0	0	0	1000
0	0	①	0.5	0	0.5	0	0	0	750
0	0	3	2	0	0	1	0	0	2400
0	0	1	1	0	0	0	-1	1	800

Subtract the transformed pivot row from the first objective row and add it to the second objective row.

A	I	x	y	s_1	s_2	s_3	s_4	a_1	RHS
1	0	0	0.5	0	-0.5	0	-1	0	50
0	1	0	-0.3	0	0.5	0	0	0	750
0	0	1	1	1	0	0	0	0	1000
0	0	①	0.5	0	0.5	0	0	0	750
0	0	3	2	0	0	1	0	0	2400
0	0	1	1	0	0	0	-1	1	800

Use the transformed pivot row to change the constraints (subtract it from row 3, subtract 3 × the transformed pivot row from row 5 and subtract it from row 6). The tableau at the end of the first pivot is shown below.

A	I	x	y	s_1	s_2	s_3	s_4	a_1	RHS
1	0	0	0.5	0	-0.5	0	-1	0	50
0	1	0	-0.3	0	0.5	0	0	0	750
0	0	0	0.5	1	-0.5	0	0	0	250
0	0	①	0.5	0	0.5	0	0	0	750
0	0	0	0.5	0	-1.5	1	0	0	150
0	0	0	0.5	0	-0.5	0	-1	1	50

This is the end of the first pivot. The obvious solution is given by putting the non-basic variables equal to zero and the basic variables equal to the right-hand side values corresponding to their 1s.

$$y = s_2 = s_4 = 0$$
$$x = 750 \quad s_1 = 250$$
$$s_3 = 150 \quad a_1 = 50$$
$$I = 750 \quad A = 50$$

The equations are:

$$A = 50 - 0.5y + 0.5s_2 + s_4$$
$$I = 750 + 0.3y - 0.5s_2$$
$$0.5y + s_1 - 0.5s_2 \qquad\qquad = 250$$
$$x + 0.5y \quad + 0.5s_2 \qquad\qquad = 750$$
$$0.5y \quad - 1.5s_2 + s_3 \qquad = 150$$
$$0.5y \quad - 0.5s_2 \qquad - s_4 + a_1 = 50$$

This solution corresponds to the point $(750, 0)$ on the graph.

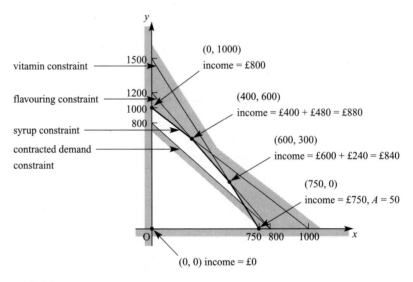

vitamin constraint

flavouring constraint

syrup constraint

contracted demand
constraint

1500

1200
1000
800

(0, 1000)
income = £800

(400, 600)
income = £400 + £480 = £880

(600, 300)
income = £600 + £240 = £840

(750, 0)
income = £750, $A = 50$

750 800   1000

(0, 0) income = £0

Figure 1.12

Since $A = 50$, the solution to this extended problem does not represent a feasible solution to the original extended problem. The fact that $A = 50$ is equivalent to $x + y = 750$, which is 50 less than the minimum allowed. The extent to which the current 'solution' is infeasible is 50.

However, further improvement is possible since the coefficient of y in the first objective row is positive. So the pivot column for the next iteration will be the y column.

Note

It may be the case that no further improvement is possible even though $A > 0$. In this case you would conclude that there is no feasible solution to the original problem, which can be an outcome.

Next you apply the ratio test to find the pivot element. This time it compares $\frac{250}{0.5} = 500$, $\frac{750}{0.5} = 1500$, $\frac{150}{0.5} = 300$ and $\frac{50}{0.5} = 100$. (Note that you do not need to look at 750 divided by -0.3.) Thus the pivot element is the 0.5 in the last row. You can then make the next pivot. At the end of the second pivot you will obtain a tableau as shown below.

A	I	x	y	s_1	s_2	s_3	s_4	a_1	RHS
1	0	0	0	0	0	0	0	−1	0
0	1	0	0	0	0.2	0	−0.6	0.6	780
0	0	0	0	1	0	0	1	−1	200
0	0	1	0	0	1	0	1	−1	700
0	0	0	0	0	−1	1	1	−1	100
0	0	0	①	0	−1	0	−2	2	100

Now $A = 0$, so you have arrived at a feasible solution to the original problem.

Interpreting the tableau, the 'obvious' solution is:

$$s_2 = s_4 = a_1 = 0$$
$$x = 700 \qquad y = 100$$
$$s_1 = 200 \qquad s_3 = 100$$
$$I = 780 \qquad A = 0$$

The equations are:

$$A = 0 + a_1$$
$$I = 780 - 0.2s_2 + 0.6s_4 - 0.6a_1$$
$$s_1 \qquad\qquad + s_4 - a_1 = 200$$
$$x \qquad + s_2 \qquad + s_4 - a_1 = 700$$
$$- s_2 + s_3 + s_4 - a_1 = 100$$
$$y \qquad - s_2 \qquad - 2s_4 + 2a_1 = 100$$

This solution corresponds to the point (700, 100) on the graph.

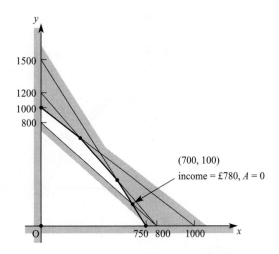

Figure 1.13

You can now dispense with your first objective row and A and a_1 columns and proceed in the usual way to a solution. This is the second stage.

I	x	y	s_1	s_2	s_3	s_4	RHS	Solution
1	0	0	0	0.2	0	−0.6	780	
0	0	0	1	0	0	1	200	(700, 100)
0	1	0	0	1	0	1	700	$I = 780$
0	0	0	0	−1	1	1	100	
0	0	1	0	−1	0	−2	100	slacks: 200, 0, 100, 0
1	0	0	0	−0.4	0.6	0	840	
0	0	0	1	1	−1	0	100	(600, 300)
0	1	0	0	2	−1	0	600	$I = 840$
0	0	0	0	−1	1	1	100	
0	0	1	0	−3	2	0	300	slacks: 100, 0, 0, 100
1	0	0	0.4	0	0.2	0	880	
0	0	0	1	1	−1	0	100	(400, 600)
0	1	0	−2	0	1	0	400	$I = 880$
0	0	0	1	0	0	1	200	
0	0	1	3	0	−1	0	600	slacks: 0, 100, 0, 200

The trajectory of the two-stage method moves from infeasible vertex to infeasible vertex in the first stage, but terminates in a feasible vertex. In the second stage the trajectory moves from feasible vertex to feasible vertex, terminating at the optimal vertex.

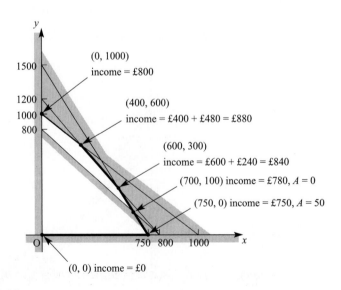

Figure 1.14

1 Use two-stage simplex to solve the modified drinks production problem. Start by using the y column as the first pivot column.

$$\text{Maximise} \quad I = x + 0.8y$$
$$\text{subject to} \quad x + y \leqslant 1000$$
$$2x + y \leqslant 1500$$
$$3x + 2y \leqslant 2400$$
$$x + y \geqslant 800$$

2 Use two-stage simplex to solve the following LP.

$$\text{Maximise} \quad P = 16x + 24y$$
$$\text{subject to} \quad 2x + 3y \leqslant 24$$
$$2x + y \leqslant 16$$
$$y \leqslant 6, \, y \geqslant 2, \, x \geqslant 0, \, y \geqslant 0$$

3 Use two-stage simplex to solve the following LP.

$$\text{Maximise} \quad P = 9x + 10y + 6z$$
$$\text{subject to} \quad 2x + 3y + 4z \leqslant 3$$
$$6x + 6y + 2z \leqslant 8$$
$$x + y + z \geqslant 1$$
$$x \geqslant 0, \, y \geqslant 0, \, z \geqslant 0$$

4 Use two-stage simplex to solve the following LP.

$$\text{Maximise} \quad P = 3w + 2x$$
$$\text{subject to} \quad w + x + y + z \leqslant 150$$
$$2w + x + 3y + 4z \leqslant 200$$
$$y + z \geqslant 50, \, w \geqslant 0, \, x \geqslant 0, \, y \geqslant 0, \, z \geqslant 0$$

5 Use two-stage simplex to solve the following LP.

$$\text{Maximise} \quad P = 3w + 2x$$
$$\text{subject to} \quad w + x + y + z \leqslant 150$$
$$2w + x + 3y + 4z \leqslant 200$$
$$w \geqslant x \quad \text{(rewrite this as } x - w \leqslant 0 \text{ to avoid the need for an}$$
$$\text{artificial variable)}$$
$$y + z \geqslant 50, \, w \geqslant 0, \, x \geqslant 0, \, y \geqslant 0, \, z \geqslant 0$$

What happens to two-stage simplex when there is no feasible solution to the LP? Investigate this with the following problem.

$$\text{Maximise} \quad P = 3x + 2y$$
$$\text{subject to} \quad 2x + 5y \leqslant 300$$
$$3x + 7y \geqslant 441$$
$$y \geqslant 20, \, x \geqslant 0, \, y \geqslant 0$$

In tackling this activity you should have introduced two artificial variables. You should have found that the minimum value that their sum can take is 1. This shows that there is no feasible solution to the problem.

The big M method

This is an alternative method for solving LP problems which include $\geqslant$ constraints. You will use it to solve the modified drinks production problem:

$$\text{Maximise} \quad I = x + 0.8y$$
$$\text{subject to} \quad x + y \leqslant 1000$$
$$2x + y \leqslant 1500$$
$$3x + 2y \leqslant 2400$$
$$x + y \geqslant 800$$

An artificial variable a_1 is introduced as before, but instead of introducing a new objective the original objective is modified to $I = x + 0.8y - Ma_1$ where M stands for some arbitrary large number.

> **Note**
>
> Again, in cases in which there is more than one $\geqslant$ constraint this new objective will be of the form $A = a_1 + a_2 + \ldots + a_n$.

Again this needs to be rewritten so as not to include a_1. You can do this by using $x + y - s_4 + a_1 = 800$, which gives $a_1 = 800 - x - y + s_4$.

So
$$I = x + 0.8y - M(800 - x - y + s_4)$$
$$= (1 + M)x + (0.8 + M)y - Ms_4 - 800M$$

Next you need to construct the initial tableau for the problem, remembering that M stands for a big number. You can imagine that M is 1000, but do not write that in for M. The initial tableau is shown below.

I	x	y	s_1	s_2	s_3	s_4	a_1	RHS
1	$-(1 + M)$	$-(0.8 + M)$	0	0	0	M	0	$-800M$
0	1	1	1	0	0	0	0	1000
0	2	1	0	1	0	0	0	1500
0	3	2	0	0	1	0	0	2400
0	1	1	0	0	0	-1	1	800

Since M is large and positive $-(1 + M)$ and $-(0.8 + M)$ are both negative, so you can improve the solution by increasing x or y.

By convention you will choose to increase x, since this is the most negative. You then apply the ratio test to find the pivot element. The ratios are as before, namely $\frac{1000}{1} = 1000$, $\frac{1500}{2} = 750$, $\frac{2400}{3} = 800$ and $\frac{800}{1} = 800$. So the pivot element is the 2 in the second constraint row.

When you make the pivot, remember that M stands for a number and follow through the algebra with M.

Scale the pivot row.

I	x	y	s_1	s_2	s_3	s_4	a_1	RHS
1	$-(1 + M)$	$-(0.8 + M)$	0	0	0	M	0	$-800M$
0	1	1	1	0	0	0	0	1000
0	①	0.5	0	0.5	0	0	0	750
0	3	2	0	0	1	0	0	2400
0	1	1	0	0	0	−1	1	800

Add $(1 + M) \times$ the pivot row to the objective row.

I	x	y	s_1	s_2	s_3	s_4	a_1	RHS
1	0	$-0.3 - 0.5M$	0	$0.5 + 0.5M$	0	M	0	$750 - 50M$
0	1	1	1	0	0	0	0	1000
0	①	0.5	0	0.5	0	0	0	750
0	3	2	0	0	1	0	0	2400
0	1	1	0	0	0	−1	1	800

Deal with the constraint rows as before.

I	x	y	s_1	s_2	s_3	s_4	a_1	RHS
1	0	$-0.3 - 0.5M$	0	$0.5 + 0.5M$	0	M	0	$750 - 50M$
0	0	0.5	1	−0.5	0	0	0	250
0	1	0.5	0	0.5	0	0	0	750
0	0	0.5	0	−1.5	1	0	0	150
0	0	0.5	0	−0.5	0	−1	1	50

There is still a negative entry in the objective row, in the y column, so another iteration is needed. Performing the ratio test gives the ratios $\frac{250}{0.5} = 500$, $\frac{750}{0.5} = 1500$, $\frac{150}{0.5} = 300$ and $\frac{50}{0.5} = 100$. So the pivot element is the entry in the last row of the y column. The next step is to make the pivot.

Scale the pivot row.

I	x	y	s_1	s_2	s_3	s_4	a_1	RHS
1	0	$-0.3 - 0.5M$	0	$0.5 + 0.5M$	0	M	0	$750 - 50M$
0	0	0.5	1	−0.5	0	0	0	250
0	1	0.5	0	0.5	0	0	0	750
0	0	0.5	0	−1.5	1	0	0	150
0	0	①	0	−1	0	−2	2	100

Deal with the objective row.

I	x	y	s_1	s_2	s_3	s_4	a_1	RHS
1	0	0	0	0.2	0	−0.6	0.6 + M	780
0	0	0.5	1	−0.5	0	0	0	250
0	1	0.5	0	0.5	0	0	0	750
0	0	0.5	0	−1.5	1	0	0	150
0	0	①	0	−1	0	−2	2	100

Deal with the constraint rows.

I	x	y	s_1	s_2	s_3	s_4	a_1	RHS
1	0	0	0	0.2	0	−0.6	0.6 + M	780
0	0	0	1	0	0	1	−1	200
0	1	0	0	1	0	1	−1	700
0	0	0	0	−1	1	1	−1	100
0	0	①	0	−1	0	−2	2	100

The M now only appears in the a_1 column. Since its presence means that a_1 will never be basic, the value of a_1 will always be zero and so the a_1 column can be deleted. The tableau is then exactly the same as at the completion of the first stage of the two-stage method (page 21), and the solution follows as before.

EXERCISE 1C

1 Use the big M method to solve the modified drinks production problem. Start by using the y column as the first pivot column.

Maximise $I = x + 0.8y$
subject to $x + y \leqslant 1000$, $2x + y \leqslant 1500$, $3x + 2y \leqslant 2400$, $x + y \geqslant 800$

2 Use the big M method to solve the following LP.

Maximise $P = 16x + 24y$
subject to $2x + 3y \leqslant 24$, $2x + y \leqslant 16$, $y \leqslant 6$, $y \geqslant 2$, $x \geqslant 0$, $y \geqslant 0$

3 Use the big M method to solve the following LP.

Maximise $P = 9x + 10y + 6z$
subject to $2x + 3y + 4z \leqslant 3$, $6x + 6y + 2z \leqslant 8$, $x + y + z \geqslant 1$,
$x \geqslant 0$, $y \geqslant 0$, $z \geqslant 0$

4 Use the big M method to solve the following LP.

Maximise $P = 3w + 2x$
subject to $w + x + y + z \leqslant 150$, $2w + x + 3y + 4z \leqslant 200$, $y + z \geqslant 50$,
$w \geqslant 0$, $x \geqslant 0$, $y \geqslant 0$, $z \geqslant 0$

5 Use the big M method to solve the following LP.

Maximise $P = 3w + 2x$
subject to $w + x + y + z \leqslant 150$, $2w + x + 3y + 4z \leqslant 200$,
$w \geqslant x$ (rewrite this as $x - w \leqslant 0$), $y + z \geqslant 50$,
$w \geqslant 0$, $x \geqslant 0$, $y \geqslant 0$, $z \geqslant 0$

EXAMPLE 1.1

Solve the following LP.

Maximise $P = x + y$

subject to $2x + y \leqslant 16$

$2x + 3y \leqslant 24$

$y \leqslant 6$

(i) graphically

(ii) using the simplex algorithm.

SOLUTION

(i)

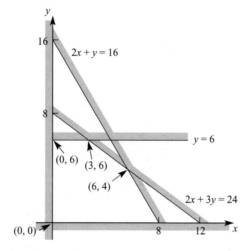

Figure 1.15

(ii)

P	x	y	s_1	s_2	s_3	RHS	
1	−1	−1	0	0	0	0	
0	2	1	1	0	0	16	$(0, 0)$
0	2	3	0	1	0	24	$P = 0$
0	0	①	0	0	1	6	
1	−1	0	0	0	1	6	
0	2	0	1	0	−1	10	$(0, 6)$
0	②	0	0	1	−3	6	$P = 6$
0	0	1	0	0	1	6	There was already a zero in the pivot column so you don't change this row
1	0	0	0	0.5	0.5	9	
0	0	0	1	−1	②	4	$(3, 6)$
0	1	0	0	0.5	−1.5	3	$P = 9$
0	0	1	0	0	1	6	
1	0	0	0.25	0.25	0	10	
0	0	0	0.5	−0.5	1	2	$(6, 4)$
0	1	0	0.75	−0.25	0	6	$P = 10$
0	0	1	−0.5	0.5	0	4	

EXAMPLE 1.2 Solve the following LP.

$$\text{Minimise} \quad C = x + y$$
$$\text{subject to} \quad 2x + y \geqslant 16$$
$$2x + 3y \geqslant 24$$
$$y \geqslant 3$$

(i) graphically

(ii) using two-stage simplex.

SOLUTION

(i)

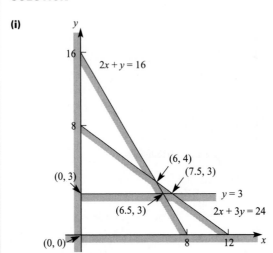

Figure 1.16

(ii)

$$2x + y - s_1 \qquad\qquad + a_1 \qquad\qquad = 16$$
$$2x + 3y \qquad - s_2 \qquad\qquad + a_2 \qquad = 24$$
$$y \qquad\qquad - s_3 \qquad\qquad + a_3 = 3$$

Adding gives

$$4x + 5y - s_1 - s_2 - s_3 + a_1 + a_2 + a_3 = 43$$
$$4x + 5y - s_1 - s_2 - s_3 + A \qquad\qquad = 43$$

A	C	x	y	s_1	s_2	s_3	a_1	a_2	a_3	RHS	
1	0	4	5	−1	−1	−1	0	0	0	43	
0	1	−1	−1	0	0	0	0	0	0	0	(0, 0) infeasible
0	0	2	1	−1	0	0	1	0	0	16	
0	0	2	3	0	−1	0	0	1	0	24	
0	0	0	①	0	0	−1	0	0	1	3	
1	0	4	0	−1	−1	4	0	0	−5	28	
0	1	−1	0	0	0	−1	0	0	1	3	(0, 3) infeasible
0	0	②	0	−1	0	1	1	0	−1	13	
0	0	2	0	0	−1	3	0	1	−3	15	
0	0	0	1	0	0	−1	0	0	1	3	
1	0	0	0	1	−1	2	−2	0	−3	2	
0	1	0	0	−0.5	0	−0.5	0.5	0	0.5	9.5	(6.5, 3) infeasible
0	0	1	0	−0.5	0	0.5	0.5	0	−0.5	6.5	
0	0	0	0	①	−1	2	−1	1	−2	2	
0	0	0	1	0	0	−1	0	0	1	3	
1	0	0	0	0	0	0	−1	−1	−1	0	
0	1	0	0	0	−0.5	0.5	0	0.5	−0.5	10.5	(7.5, 3) feasible $C = 10.5$
0	0	1	0	0	−0.5	1.5	0	0.5	−1.5	7.5	
0	0	0	0	1	−1	②	−1	1	−2	2	
0	0	0	1	0	0	−1	0	0	1	3	
1		0	0	−0.25	−0.25	0				10	
0		1	0	−0.75	0.25	0				6	(6, 4) $C = 10$
0		0	0	0.5	−0.5	1				1	
0		0	1	0.5	−0.5	0				4	

> You could have chosen s_3 as the pivot column but, as the pivot element in the s_1 column is 1, this gives less work

---

EXAMPLE 1.3

Solve the following LP.

Minimise $C = x + y$
subject to $2x + y \geqslant 16$
$2x + 3y \geqslant 24$
$y \geqslant 3$

(i) graphically
(ii) using the big M method.

SOLUTION

(i)

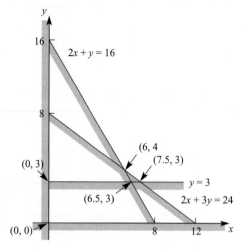

Figure 1.17

(ii)
$$C = x + y + M(a_1 + a_2 + a_3)$$
$$= x + y + M(16 - 2x - y + s_1 + 24 - 2x - 3y + s_2 + 3 - y + s_3)$$
$$= x(1 - 4M) + y(1 - 5M) + Ms_1 + Ms_2 + Ms_3 + 43M$$
$$\text{So } C + (-1 + 4M)x + (-1 + 5M)y - Ms_1 - Ms_2 - Ms_3 = 43M$$

C	x	y	s_1	s_2	s_3	RHS	
1	$-1 + 4M$	$-1 + 5M$	$-M$	$-M$	$-M$	$43M$	
0	2	1	-1	0	0	16	$(0, 0)$
0	2	3	0	-1	0	24	infeasible
0	0	①	0	0	-1	3	
1	$-1 + 4M$	0	$-M$	$-M$	$-1 + 4M$	$3 + 28M$	
0	②	0	-1	0	1	13	$(0, 3)$
0	2	0	0	-1	3	15	infeasible
0	0	1	0	0	-1	3	
1	0	0	$-0.5 + M$	$-M$	$-0.5 + 2M$	$9.5 + 2M$	
0	1	0	-0.5	0	0.5	6.5	$(6.5, 3)$
0	0	0	1	-1	2	2	infeasible
0	0	1	0	0	-1	3	
1	0	0	0	-0.5	0.5	10.5	
0	1	0	0	-0.5	1.5	7.5	$(7.5, 3)$
0	0	0	1	-1	2	2	feasible
0	0	1	0	0	-1	3	$C = 10.5$
1	0	0	-0.25	-0.25	0	10	
0	1	0	-0.75	0.25	0	6	$(6, 4)$
0	0	0	0.5	-0.5	1	1	$C = 10$
0	0	1	0.5	-0.5	0	4	

Equalities and practicalities

You have seen how the simplex method was developed to deal with $\leqslant$ inequalities. You have seen how, by applying itself to the problem, it can deal with $\geqslant$ inequalities. What about equalities?

Suppose that you need to impose the condition $x + y = 80$. To deal with this using simplex it is simply replaced by two inequality constraints:

$$x + y \leqslant 80$$
$$\text{and} \quad x + y \geqslant 80$$

Of course, this involves dealing with a $\geqslant$ inequality. It is best in this context to use the big M method, since two-stage simplex can lead to a slight technical complication known as a degenerate basis. This is easy to deal with but, by using big M, it will not arise.

In practice, modifications to the basic simplex method (such as allowing a variable to be negative, dealing with $\geqslant$ inequalities and dealing with equalities) are coded into software packages, and so are dealt with automatically. This is a good example of mathematics underpinning systems, without being visible.

Finally, what about $<$ or $>$ constraints? Since you are dealing with linear programming the mathematical assumption is that the variables can take real number values. In practice LP problems are solved on computer systems, which operate with rational numbers.

Suppose a crude computer operates with four-digit accuracy and suppose you wished to model the constraint $x < 4$. Would you allow the machine to accept values of x up to 3.999 or values up to $x = 4$? If you chose the former then all values greater than 3.999 and less than 4 would be mistakenly rejected; if the latter then only the single value $x = 4$ would be mistakenly accepted. Thus LP computer packages generally draw no distinction between $x \leqslant 4$ and $x < 4$.

EXAMPLE 1.4

Solve the following LP.

$$\text{Maximise} \quad P = 2x + y$$
$$\text{subject to} \quad x + y \leqslant 2,$$
$$y = 1$$

SOLUTION

It is not difficult to see that the answer to this is $P = 3$ at $x = 1$ and $y = 1$, but it is instructive to see how the method for dealing with equalities works.

Note that the objective is constructed from
$$P = 2x + y - Ma$$
$$= 2x + y - M(1 - y + s_3).$$

P	x	y	s_1	s_2	s_3	a	RHS
1	-2	-(1 + M)	0	0	M	0	-M
0	1	1	1	0	0	0	2
0	0	①	0	1	0	0	1
0	0	1	0	0	-1	1	1
1	-2	0	0	1 + M	M	0	1
0	①	0	1	-1	0	0	1
0	0	1	0	1	0	0	1
0	0	0	0	-1	-1	1	0
1	0	0	2	M - 1	M	0	3
0	1	0	1	-1	0	0	1
0	0	1	0	1	0	0	1
0	0	0	0	-1	-1	1	0

EXERCISE 1D

1 (i) Use two-stage simplex to solve the following LP. Relate each tableau to a point on the graph of the feasible region.

Minimise $y + 2x$

Subject to $x + y \geqslant 20$, $2x + 5y \geqslant 70$, $7x + 3y \geqslant 84$

(ii) Solve the problem using the big M method.

2 (i) Solve the following LP.

Maximise $3x + y + z$

Subject to $x + y + z \leqslant 3$, $x + y + 2z \leqslant 4$, $x \geqslant 1$, $y \leqslant 2$

(ii) Solve the following LP.

Maximise $3x + y + z$

Subject to $x + y + z \leqslant 3$, $x + y + 2z = 4$, $x \geqslant 1$, $y \leqslant 2$

3 (i) Solve the following LP.

Minimise $3x + 5y$

Subject to $6x + 5y \geqslant 60$, $2x + 3y \geqslant 24$, $x + y \leqslant 12$

(a) graphically   **(b)** using simplex.

(ii) By considering your graph say what the consequence is of changing the final constraint to $x + y \leqslant 10$.

(iii) Use simplex to attempt to solve the following LP.

Maximise $3x + 5y$

Subject to $6x + 5y \geqslant 60$, $2x + 3y \geqslant 24$, $x + y \leqslant 10$

4 (i) Solve the following LP.

  Minimise $2x + y$
  Subject to $x + y \geqslant 47, x + 10y \geqslant 110, 10x + y \geqslant 110$

  (a) graphically   **(b)** using simplex.

 (ii) Solve the following LP.

  Minimise $2x + y$
  Subject to $x + y \geqslant 20, x + 10y \geqslant 110, 10x + y \geqslant 110$

  (a) graphically   **(b)** using simplex.

 (iii) Solve the following LP.

  Minimise $2x + y$
  Subject to $x + y \geqslant 11, x + 10y \geqslant 110, 10x + y \geqslant 110$

  (a) graphically   **(b)** using simplex.

5 Attempt to solve the following LP.

  Maximise $x + 2y$
  Subject to $x + y \geqslant 6, x \geqslant 2, y \geqslant 2$

  (i) graphically   **(ii)** using simplex.

6 A production unit makes two types of product, X and Y. Production levels are measured in tonnes and are constrained by the availability of finance, staff and storage space. Requirements for, and daily availabilities of, each of these resources are summarised in the table.

	Finance (£)	Staff time (hours)	Storage space (m³)
Requirement per tonne for product X	200	8	1
Requirement per tonne for product Y	100	8	3
Resources available per day	1000	48	15

The profit on these products is £160 per tonne for X and £120 per tonne for Y.

(i) Express the three resource constraints as inequalities and write two further inequalities indicating that production levels are non-negative.

(ii) Given that the objective is to maximise profit, state the objective function.

(iii) Illustrate the inequalities graphically, and use your graph to find the best daily production plan.

(iv) Set up an initial tableau for the problem and use the simplex algorithm to solve it. Relate each stage of the tableau to its corresponding point on your graph.

[Oxford]

7 In an executive initiative course, participants are asked to travel as far as possible in three hours using a combination of moped, car and lorry. The moped can be carried in the car and the car can be carried on the lorry.

The moped travels at 20 miles per hour (mph) with a petrol consumption of 60 miles per gallon (mpg). The car travels at 40 mph with a petrol consumption of 40 mpg. The lorry travels at 30 mph with a petrol consumption of 20 mpg.

2.5 gallons of petrol are available.

The moped must not be used for more than 55 miles, and a total of no more than 55 miles must be covered using the car and/or lorry.

(i) Formulate the above specifications as a linear programming problem, stating clearly your choice of variables and your objective function.
(ii) Set up a simplex tableau to solve the problem.
(iii) Perform two iterations of the simplex algorithm.
(iv) Is the tableau resulting from part (iii) optimal?

What strategy is indicated by the tableau, and how much time and petrol would be used?

[Oxford]

8 A diet planning problem involves mixing two foodstuffs, X and Y, to satisfy three dietary constraints at minimum cost, C. The amounts of X and Y used are x kg and y kg respectively. The cost is £1 per kg for each.

The problem is formulated as a linear program:

Minimise $C = x + y$
subject to $2x + y \geqslant 8$
$3x + 2y \geqslant 14$
$x + 2y \geqslant 6$
$x \geqslant 0$
$y \geqslant 0$

(i) Draw a graph to illustrate the inequalities and the feasible region.
(ii) The inequalities are converted into equalities by introducing surplus variables, s_1, s_2 and s_3, and are laid out in a tableau as shown below.

C	x	y	s_1	s_2	s_3	RHS
1	−1	−1	0	0	0	0
0	2	1	−1	0	0	8
0	3	2	0	−1	0	14
0	1	2	0	0	−1	6

Explain why it is not possible to apply the simplex algorithm to this tableau.

(iii) A modification to the simplex algorithm is applied to produce a feasible solution to the above problem. Expressed in tableau form the solution is as shown below.

C	x	y	s_1	s_2	s_3	RHS
1	1	0	−1	0	0	8
0	2	1	−1	0	0	8
0	1	0	−2	1	0	2
0	3	0	−2	0	1	10

   (a) Interpret this tableau by stating the values of the variables x and y and the cost C, and show that the constraints are satisfied.

   (b) What are the values taken by s_1, s_2 and s_3 in this solution? How do these values relate to the constraints?

(iv) Starting from the tableau in part (iii) two iterations of the simplex algorithm are needed to achieve the optimal solution. With this formulation you will choose to pivot on a column (other than the C column) that has a positive value in the first row of the table. Thus the first iteration will involve increasing the value of x. Perform the two iterations, interpret the result and mark the point to which it relates on your graph.

[Oxford]

9 A linear programming problem is specified as:

Maximise $\quad P = 6x + 4y$

subject to $\quad x + 2y \leqslant 14$
$\qquad\qquad x + y \leqslant 8$
$\qquad\qquad 2x + y \leqslant 11$
$\qquad\qquad 3x + y \leqslant 15$
$\qquad\qquad x \geqslant 0, y \geqslant 0$

(i) Solve this problem graphically.

To solve the problem using the simplex method four slack variables have to be introduced, changing the first four inequalities to:

$x + 2y + s_1 = 14$
$x + y + s_2 = 8$
$2x + y + s_3 = 11$
$3x + y + s_4 = 15$

(ii) At an intermediate stage in the application of the simplex method the slack variables take the values $s_1 = 4$, $s_2 = 1$, $s_3 = 0$, $s_4 = 0$.

   (a) To what point on your graph from part (i) does this correspond?

   (b) Express P in terms of s_3 and s_4.

   Explain why your expression shows that the solution is not optimal.

(iii) Complete the application of the simplex method.
    (a) Give the values of the slack variables at the optimal point.
    (b) For each inequality say what information is given by the value of its slack variable.

[AEB]

10 A sum of money is to be invested in three shares, A, B and C. The proportions to be invested in each are a, b and c. The returns from each share are 0.11, 0.15 and 0.08 respectively. A measure of the risk entailed in investing in the shares is given by the expression $a + 1.2b + 0.9c$. The risk is to be minimised subject to the investment achieving a total return of at least 0.10 (that is, 10%).

(i) State the total return from the investment in terms of a, b and c. If all of the money is to be invested show that the total return can be written as $0.08 + 0.03a + 0.07b$ and give an expression for the risk in terms of a and b only.

The investment manager imposes three extra constraints:
(a) no more than half of the money is to be invested in A
(b) no more than a quarter of the money is to be invested in B
(c) at least a third of the money is to be invested in C.

(ii) Express these three new constraints as inequalities involving a and b only.
(iii) Use a graphical approach to solve the problem of choosing a and b (and hence c) to minimise the risk, subject to all four constraints.
(iv) The simplex method provides a way of moving from a feasible solution to an LP problem to an improved feasible solution. What difficulty has to be overcome in applying the method to a problem involving inequalities of the type $0.03a + 0.07b \geqslant 0.02$?

[AEB]

11 (i) Solve the following linear programming problem graphically.

    Maximise $P = x + 2y$
    subject to $3x + 17y \leqslant 170$
                  $7x + 8y \leqslant 175$
                  $y \leqslant 9$
                  $x \leqslant 20$

(ii) Starting from the initial feasible solution $x = y = 0$, apply two iterations of the simplex method to the linear programming problem given in part (i). Identify the points on your graph corresponding to the results of your iterations of the simplex method.
(iii) Give the values of the slack variables in the *final* simplex solution to the problem.

[AEB, adapted]

12 A company producing dining tables and chairs requires a production plan for the next month. The company has £10 000 budgeted to buy materials. Chairs each require £20 of materials and tables £100. Tables each take 15 hours to complete and chairs each take 4 hours. There are 1950 hours available per month. The company sells chairs at £80 each and tables at £350. The production plan is required to maximise potential income.

(i) Formulate the problem as a linear programming problem.

(ii) Use a graphical method to solve the problem.

(iii) Had the simplex method been used, slack variables would have been needed. State what those slack variables would have represented and give their values at the solution.

(iv) Comment on any practical difficulty there may be with the linear programming solution to this problem.

[AEB]

13 Three products, X, Y and Z, are to be manufactured. They all require resources A, B, C and D which are in limited supply. The table summarises these requirements in suitable units per item produced.

	A	B	C	D
X	2	0	2	4
Y	5	2	4	3
Z	4	1	2	2
Availability	60	10	70	180

Profits are £3 per item of X produced, £2 per item of Y and £5 per item of Z.

(i) Formulate a linear programming problem to maximise profit within the constraints imposed by resource availabilities.

(ii) Use the simplex algorithm to solve the problem.

(iii) An extra constraint is imposed by a contract to supply at least 5 items of Y. Show how to incorporate this constraint into an initial tableau using a surplus and an additional variable.

(iv) Explain how to use either two-stage simplex or the big M method to move to a feasible solution to the modified problem. You should show the initial tableau, including the objective function(s), and explain briefly how to proceed. You are not required to do the iterations.

[MEI]

14 A manufacturer of garden furniture produces chairs, round tables and square tables. There must be at least 4 chairs produced for each table. At least 100 round tables and 80 square tables must be produced. The costs of manufacture are £4 per chair, £10 per round table and £8 per square table.

(i) Using x, y and z to represent the numbers of chairs, round tables and square tables produced respectively, formulate as a linear program the problem of deciding how many of each item to produce at minimum cost.

(ii) The initial tableau and the final tableau for a two-stage simplex solution to the LP are shown below. Explain the structure of the initial tableau, including the variables and the two objective functions. Interpret the final tableau.

Initial tableau

Q	C	x	y	z	s_1	s_2	s_3	a_2	a_3	RHS
1	0	0	1	1	0	−1	−1	0	0	180
0	1	−4	−10	−8	0	0	0	0	0	0
0	0	−1	4	4	1	0	0	0	0	0
0	0	0	1	0	0	−1	0	1	0	100
0	0	0	0	1	0	0	−1	0	1	80

Final tableau

Q	C	x	y	z	s_1	s_2	s_3	a_2	a_3	RHS
1	0	0	0	0	0	0	0	−1	−1	0
0	1	0	0	0	−4	−26	−24	26	24	4520
0	0	1	0	0	−1	−4	−4	4	4	720
0	0	0	1	0	0	−1	0	1	0	100
0	0	0	0	1	0	0	−1	0	1	80

(iii) Chairs are sold to retailers at £8 each, round tables at £15 each and square tables at £12 each. Write down an expression in terms of x, y and z for the total profit.

(iv) The manufacturer wishes to maximise the profit, P, while spending no more than £5000 on manufacturing costs. You are given that the tableau shown below takes the solution represented by the final tableau in part **(ii)** as the starting point for this problem. Apply the simplex algorithm to this tableau to find the most profitable production plan, pivoting on the s_1 column.

P	x	y	z	s_1	s_2	s_3	s_4	RHS
1	0	0	0	−4	−21	−20	0	3700
0	1	0	0	−1	−4	−4	0	720
0	0	1	0	0	−1	0	0	100
0	0	0	1	0	0	−1	0	80
0	0	0	0	4	26	24	1	480

[MEI]

15 Three chemical products, X, Y and Z are to be made. Product X will sell at 40p per litre and costs 30p per litre to produce. Product Y will sell at 40p per litre and costs 30p per litre to produce. Product Z will sell at 40p per litre and costs 20p per litre to produce.

Three additives are used in each product. Product X uses 5 g per litre of additive A, 2 g per litre of additive B and 8 g per litre of additive C. Product Y uses 2 g per litre of additive A, 4 g per litre of additive B and 3 g per litre of additive C. Product Z uses 10 g per litre of additive A, 5 g per litre of additive B and 5 g per litre of additive C.

There are 10 kg of additive A available, 12 kg of additive B, and 8 kg of additive C.

(i) Explain how the initial feasible tableau shown below models this problem.

P	x	y	z	s_1	s_2	s_3	RHS
1	−10	−10	−20	0	0	0	0
0	5	2	10	1	0	0	10 000
0	2	4	5	0	1	0	12 000
0	8	3	5	0	0	1	8 000

(ii) Use the simplex algorithm to solve your LP, and interpret your solution.

(iii) The optimal solution involves making two of the three products. By how much would the cost of making the third product have to fall to make it worth producing, assuming that the selling price is not changed?

There is a contractual requirement to provide at least 500 litres of product X.

(iv) Show how to incorporate this constraint into the initial tableau ready for an application of the two-stage simplex method. Briefly describe how the method works. You are not required to perform the iterations.

[MEI]

16 The graph shows the feasible region for the following LP problem.

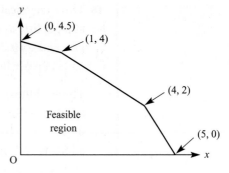

Maximise $P = \frac{1}{3}x + \frac{1}{2}y$

subject to $x + 2y \leqslant 9$

$2x + 3y \leqslant 14$

$2x + y \leqslant 10$

$x \geqslant 0$ and $y \geqslant 0$

(i) Use the graph to solve the LP problem.

(ii) Solve the problem using the simplex algorithm. Start with pivoting on the x column. Show that the final tableau is

P	x	y	s_1	s_2	s_3	RHS
1	0	0	0	$\frac{1}{6}$	0	$\frac{7}{3}$
0	0	0	1	$-\frac{3}{4}$	$\frac{1}{4}$	1
0	0	1	0	$\frac{1}{2}$	$-\frac{1}{2}$	2
0	1	0	0	$-\frac{1}{4}$	$\frac{3}{4}$	4

Interpret this solution.

(iii) From the final tableau given in part (ii), perform another pivot using the $\frac{1}{4}$ as pivot element. Comment on the result.

The simplex algorithm is applied to the problem

Maximise $Q = \frac{1}{3}x + \frac{1}{2}y + \frac{7}{12}z$

subject to $4x + 8y + 9z \leqslant 36, \ 4x + 6y + 7z \leqslant 28, \ 4x + 2y + 5z \leqslant 20,$

$x \geqslant 0, \ y \geqslant 0$ and $z \geqslant 0$

The following tableau is produced.

Q	x	y	z	s_1	s_2	s_3	RHS
1	0	0	0	0	$\frac{1}{12}$	0	$\frac{7}{3}$
0	$-\frac{8}{7}$	$\frac{2}{7}$	0	1	$-\frac{9}{7}$	0	0
0	$\frac{4}{7}$	$\frac{6}{7}$	1	0	$\frac{1}{7}$	0	4
0	$\frac{8}{7}$	$-\frac{16}{7}$	0	0	$-\frac{5}{7}$	1	0

(iv) Interpret this tableau. Say how you know that there are other optimal solutions, and find the two other optimal vertices. (Hint: Put $z = 0$ in the statement of the problem.)

[MEI]

1 The simplex algorithm provides the means for moving from one simplex tableau to another. At each stage the essential structure is preserved.

- There is one objective row and the rest are constraint rows.
- For each row there is one column containing a 1 and the remaining spaces are filled with zeros. One of these columns corresponds to the objective. The other columns correspond to variables which are called basic variables.
- The remaining columns (other than the RHS) correspond to variables known as non-basic variables.
- The entries in the RHS column are all non-negative.

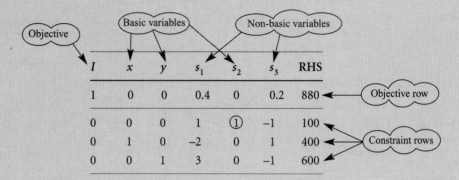

2 There are three steps involved in an iteration of the simplex algorithm:

- identifying a pivot column
- using the ratio test to identify the pivot element
- making the pivot.

3 There are two methods to accommodate $\geq$ inequalities:

- two-stage simplex
- the big M method.

2 Networks: touring and inspecting

Marry, this is the short and long of it.

The Merry Wives of Windsor, *Shakespeare*

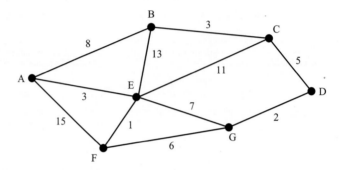

Figure 2.1

? **1** What is the least weight path from A to D in the network in figure 2.1?

2 What is the least weight path from A to each of B, C, E, F and G?

3 How many other 'least weight path' questions could you be asked about this network?

Floyd's algorithm

In *Decision Mathematics 1* you used Dijkstra's algorithm to find the least weight path from one vertex in a network to another. Dijkstra's algorithm has quadratic complexity.

Dijkstra's algorithm is a labelling algorithm and a label is assigned to a vertex when the shortest distance to that vertex is known for certain. The algorithm can be stopped when the destination vertex is labelled. If you continued until all other vertices were labelled, then you would find the shortest distances (and thence paths) from the initial vertex to each of the other vertices, at a 'cost' which is $O(n^2)$.

You will see later in this chapter that there are circumstances in which you need *all* shortest paths in a network. You could achieve this by applying Dijkstra *n* times, one for each starting vertex. This would have complexity $O(n^3)$. In this section you will meet an alternative cubic algorithm, Floyd's algorithm, which achieves the same ends.

You will be using the network in figure 2.2, which is simpler than that in figure 2.1. This will help you to see how the algorithm works. You would have no difficulty in writing down, by inspection, the ten shortest distances in this network. For larger networks, however, inspection is of no use. You must, as in other areas of decision mathematics, operate algorithmically. You will need an efficient algorithm for this task, and one which can be implemented on a computer.

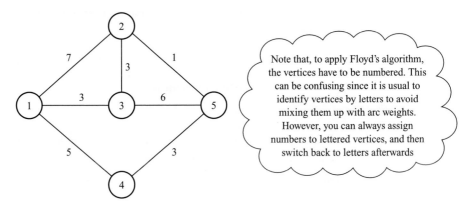

Note that, to apply Floyd's algorithm, the vertices have to be numbered. This can be confusing since it is usual to identify vertices by letters to avoid mixing them up with arc weights. However, you can always assign numbers to lettered vertices, and then switch back to letters afterwards

Figure 2.2

❓ Why are there ten shortest paths in the network in figure 2.2? How many would there have been had the network consisted of ten connected vertices? How many if there were n vertices?

The matrices

The method successively modifies two matrices.

A distance (or weight) matrix

This gives the distance from the start vertex to the destination vertex. It is symmetrical unless you are dealing with a digraph. In the first instance it shows the length of each arc in the network. Where there is no direct connection (i.e. no edge) the distance is shown as ∞. For instance, the direct distance from a vertex to itself should be shown as ∞, unless there is a loop at that vertex. You may need to replace a dash by ∞.

The initial distance matrix for the example in figure 2.2 is shown overleaf.

destination node

		1	2	3	4	5
	1	∞	7	3	5	∞
start	2	7	∞	3	∞	1
node	3	3	3	∞	∞	6
	4	5	∞	∞	∞	3
	5	∞	1	6	3	∞

A route matrix

This matrix shows the first vertex encountered en route from the start vertex to the destination vertex. The initial version of this has a column of 1s followed by a column of 2s, and so on.

The initial route matrix for the example in figure 2.2 is shown below.

destination node

		1	2	3	4	5
	1	1	2	3	4	5
start	2	1	2	3	4	5
node	3	1	2	3	4	5
	4	1	2	3	4	5
	5	1	2	3	4	5

3 in this cell indicates that the first node you encounter en route from vertex 2 to vertex 3 is 3, that is, you are taking the direct route

The algorithm

Step 1 Choose the first (or next in subsequent iterations) row and column and note the iteration number (starting at 1). Highlight all the elements in the chosen row and column.

Step 2 Compare each element in the distance matrix which is not in the chosen row or column, with the sum of the corresponding entry in the chosen row and the corresponding entry in the chosen column.

Step 3 If the calculated sum is smaller than the element then replace the element with the sum.

Step 4 If the element is replaced you need to update the corresponding entry of the route matrix. The value to replace it is found in the route matrix from the previous iteration in the same row as the entry it is replacing, but in the column given by the iteration number.

Step 5 Repeat steps 1 to 4 until all rows and columns have been chosen in step 1.

Step 6 The distance matrix will now contain the shortest distances. The route matrix shows the next vertex en route from one to another along a shortest path. It can be used to construct a complete set of shortest paths.

The first iteration

The matrices for the network in figure 2.2 are shown below. The entries in the revised tables which have been changed are printed in **bold** type.

D(0)

	1	2	3	4	5
1	∞	7	3	5	∞
2	7	∞	3	∞	1
3	3	3	∞	∞	6
4	5	∞	∞	∞	3
5	∞	1	6	3	∞

R(0)

	1	2	3	4	5
1	1	2	3	4	5
2	1	2	3	4	5
3	1	2	3	4	5
4	1	2	3	4	5
5	1	2	3	4	5

For example, ∞ has been changed to 12 because the highlighted number in its row (7) plus the highlighted number in its column (5) is lower

This entry has been changed because the corresponding distance, ∞, was changed. Its new value comes from the second row of R(0), and the first column, since this is the first iteration

D(1)

	1	2	3	4	5
1	∞	7	3	5	∞
2	7	**14**	3	**12**	1
3	3	3	**6**	**8**	6
4	5	**12**	**8**	**10**	3
5	∞	1	6	3	∞

R(1)

	1	2	3	4	5
1	1	2	3	4	5
2	1	**1**	3	**1**	5
3	1	2	**1**	**1**	5
4	1	**1**	**1**	**1**	5
5	1	2	3	4	5

The second iteration

D(1)

	1	2	3	4	5
1	∞	7	3	5	∞
2	7	14	3	12	1
3	3	3	6	8	6
4	5	12	8	10	3
5	∞	1	6	3	∞

R(1)

	1	2	3	4	5
1	1	2	3	4	5
2	1	1	3	1	5
3	1	2	1	1	5
4	1	1	1	1	5
5	1	2	3	4	5

In this cell, 4 replaces 6. It is the sum of 1, the highlighted number in 6's row, and 3, the highlighted number in 6's column. 6 represents the direct distance from vertex 5 to vertex 3; 1 is the distance from vertex 5 to vertex 2 and 3 is the distance from vertex 2 to vertex 3

As this cell in the distance matrix has been changed, the corresponding entry in this matrix has to be changed. The new value is taken from the fifth row of R(1) and the second column, since this is the second iteration

D(2)

	1	2	3	4	5
1	**14**	7	3	5	**8**
2	7	14	3	12	1
3	3	3	6	8	**4**
4	5	12	8	10	3
5	**8**	1	**4**	3	**2**

R(2)

	1	2	3	4	5
1	**2**	2	3	4	**2**
2	1	1	3	1	5
3	1	2	1	1	**2**
4	1	1	1	1	5
5	**2**	2	**2**	4	**2**

In the second iteration you have been asking if you can do better by going via vertex 2. Before the iteration the shortest path you had found from vertex 5 to vertex 3 was of length 6. It was the direct path. During the iteration this was replaced by the path $5 \rightarrow 2 \rightarrow 3$, which is of length 4. The route matrix shows that the first vertex encountered on the current best path from 5 to 3 is vertex 2.

The third iteration

D(2)

	1	2	3	4	5
1	14	7	3	5	8
2	7	14	3	12	1
3	3	3	6	8	4
4	5	12	8	10	3
5	8	1	4	3	2

R(2)

	1	2	3	4	5
1	2	2	3	4	2
2	1	1	3	1	5
3	1	2	1	1	2
4	1	1	1	1	5
5	2	2	2	4	2

D(3)

	1	2	3	4	5
1	6	6	3	5	7
2	6	6	3	11	1
3	3	3	6	8	4
4	5	11	8	10	3
5	7	1	4	3	2

R(3)

	1	2	3	4	5
1	3	3	3	4	3
2	3	3	3	3	5
3	1	2	1	1	2
4	1	1	1	1	5
5	2	2	2	4	2

This is showing that there is a path of length 7 from vertex 5 to vertex 1. The route matrix shows that the first vertex encountered on that path is vertex 2. To find the next vertex look at the second row and first column of R(3). This gives the first vertex en route from vertex 2 to 1

These two might have caught you out. They come from the fourth and fifth rows of R(2), and from the third column, since this is the third iteration

As an example, the best path you have found so far from vertex 5 to vertex 1 is $5 \rightarrow 2 \rightarrow 3 \rightarrow 1$ and is of length 7.

The fourth iteration

D(3)

	1	2	3	4	5
1	6	6	3	5	7
2	6	6	3	11	1
3	3	3	6	8	4
4	5	11	8	10	3
5	7	1	4	3	2

R(3)

	1	2	3	4	5
1	3	3	3	4	3
2	3	3	3	3	5
3	1	2	1	1	2
4	1	1	1	1	5
5	2	2	2	4	2

D(4)

	1	2	3	4	5
1	6	6	3	5	7
2	6	6	3	11	1
3	3	3	6	8	4
4	5	11	8	10	3
5	7	1	4	3	2

R(4)

	1	2	3	4	5
1	3	3	3	4	3
2	3	3	3	3	5
3	1	2	1	1	2
4	1	1	1	1	5
5	2	2	2	4	2

There were no changes during this iteration.

The final iteration

D(4)

	1	2	3	4	5
1	6	6	3	5	7
2	6	6	3	11	1
3	3	3	6	8	4
4	5	11	8	10	3
5	7	1	4	3	2

R(4)

	1	2	3	4	5
1	3	3	3	4	3
2	3	3	3	3	5
3	1	2	1	1	2
4	1	1	1	1	5
5	2	2	2	4	2

D(5)

	1	2	3	4	5
1	6	6	3	5	7
2	6	2	3	4	1
3	3	3	6	7	4
4	5	4	7	6	3
5	7	1	4	3	2

R(5)

	1	2	3	4	5
1	3	3	3	4	3
2	3	5	3	5	5
3	1	2	1	2	2
4	1	5	5	5	5
5	2	2	2	4	2

This is the end of the algorithm. The D(5) matrix contains all of the shortest distances. The shortest paths can be constructed from the R(5) matrix.

Interpreting the matrices

Look at the route from vertex 1 to vertex 5 as an example.

D(5)

	1	2	3	4	5
1	6	6	3	5	7
2	6	2	3	4	1
3	3	3	6	7	4
4	5	4	7	6	3
5	7	1	4	3	2

This entry in the distance matrix tells you that the shortest distance from vertex 1 to vertex 5 is 7

The route is found by examining the route matrix.

R(5)

	1	2	3	4	5
1	3	3	3	4	3
2	3	5	3	5	5
3	1	2	1	2	2
4	1	5	5	5	5
5	2	2	2	4	2

The first vertex on the shortest path is vertex 3

The next vertex on the path is found by looking for the first vertex on the shortest path from 3 to 5. This is vertex 2

Finally, the first vertex on the shortest path from 2 to 5 is, in fact, 5 indicating that you have reached your destination

The complete path is $1 \rightarrow 3 \rightarrow 2 \rightarrow 5$ and is of length 7.

? What is the shortest distance and the shortest path from vertex 3 to vertex 4?

Pseudo-code

The algorithm can be written very succinctly in pseudo-code, showing how efficiently it can be coded for computer implementation. It is, however, difficult to follow written in this form. (You do not need to know this pseudo-code.)

For $k = 1, 2, ..., n$
    For $i = 1, 2, ..., n$
        For $j = 1, 2, ..., n$

$D^{(k)}(i, j)$ means the entry in the ith row and the jth column of the kth distance matrix

$$D^{(k)}(i, j) = \min\{D^{(k-1)}(i, j), D^{(k-1)}(i, k) + D^{(k-1)}(k, j)\}$$
If $D^{(k)}(i, j) = D^{(k-1)}(i, j)$ then $R^{(k)}(i, j) = R^{(k-1)}(i, j)$
If $D^{(k)}(i, j) \neq D^{(k-1)}(i, j)$ then $R^{(k)}(i, j) = R^{(k-1)}(i, k)$

        Next j
    Next i
Next k

The nested loop structure does show that the complexity of the algorithm is $O(n^3)$. The inside loop ($j = 1, 2, ..., n$) is executed n times for each execution of the middle loop, ($i = 1, 2, ..., n$). The three core statements are executed n^2 times for each execution of the outside loop. The outside loop is executed n times ($k = 1, 2, ..., n$), so the core statements are executed n^3 times.

EXERCISE 2A

1 Apply Floyd's algorithm to the networks represented by the following matrices.

(i)

	1	2	3	4
1	–	9	–	3
2	9	–	2	–
3	–	2	–	2
4	3	–	2	–

(ii)

	1	2	3	4	5	6	7	8
1	–	9	–	3	–	–	–	–
2	9	–	2	–	7	–	–	–
3	–	2	–	2	4	8	6	–
4	3	–	2	–	–	–	5	–
5	–	7	4	–	–	10	–	9
6	–	–	8	–	10	–	7	12
7	–	–	6	5	–	7	–	10
8	–	–	–	–	9	12	10	–

2 Apply Floyd's algorithm to the following networks.

(i)

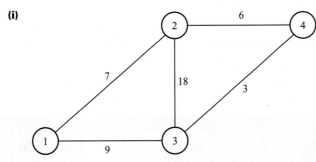

(ii)

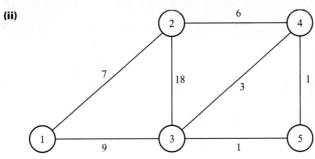

(iii)

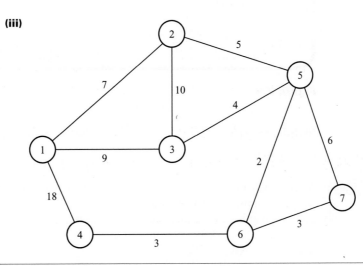

⚠ Question 1 part (ii) and question 2 part (iii) involve a lot of work.

The travelling salesperson problem

The advertisement in figure 2.3 shows the sequence of venues in a White Light concert tour. With concerts almost every day and a lot of travelling in between, this will have been a tiring schedule.

? Using your knowledge of geography and assuming that the concert halls were available, could the sequence of towns and cities be arranged in a better way to cut down on the total distance travelled?

Problems like those of the concert tour are called *travelling salesperson problems* after the classic situation of a sales representative who has to make a round trip, starting at home, visiting a sequence of towns and finally returning home. The problem is to choose a route that minimises the total distance travelled.

Figure 2.3

July			
Sat	11th	LONDON	Hammersmith Odeon
Sun	12th	OXFORD	Apollo
Tue	14th	MANCHESTER	Apollo
Wed	15th	LIVERPOOL	Royal Court
Fri	17th	ABERDEEN	Music Hall
Sat	18th	GLASGOW	Concert Hall

Mon	20th	NEWCASTLE	City Hall
Tue	21st	SHEFFIELD	City Hall
Wed	22nd	NOTTINGHAM	Royal Centre
Fri	24th	BIRMINGHAM	Town Hall
Sun	26th	BRIGHTON	Dome
Tue	28th	BRISTOL	Colston Hall
Thu	30th	LONDON	Hammersmith Odeon

All tickets available from 28th March.   Call LONDON CREDIT CARD HOTLINE

In order to get a feel for the problem, try the following exercise before reading any further.

EXERCISE 2B

A bread delivery van delivers each day to the towns shown below. Find the route with the smallest distance which starts at the bakery in Worcester, passes through each town at least once and returns to Worcester.

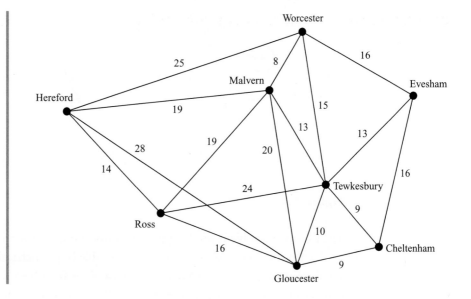

The shortest distance is given in the answer section at the back of the book. How close did you get? Did you adopt any particular strategy?

The solution for a small network can be found by trial and error, but it is necessary to develop algorithms that can be applied in a systematic way to deal with larger problems.

Start the investigation by looking at a small problem with five towns, A, B, C, D and E. Assume A is the home town. The table below shows the shortest distance from each town to all the others.

	A	B	C	D	E
A	–	8	10	11	7
B	8	–	17	5	14
C	10	17	–	12	9
D	11	5	12	–	11
E	7	14	9	11	–

As the salesperson leaves his home town A, he has the choice of going to any one of the remaining four towns: assume he goes to C. From town C he can travel to any one of the remaining three, B, D or E, of which he chooses E. Suppose his next choice is B, he must then choose D and finally return to A. This is shown in figure 2.4, overleaf.

The ordering of the towns A → C → E → B → D → A is sometimes called a *tour*. The distance between two towns, say A and C, is denoted d_{AC}.

For this tour the total distance is

$$d_{AC} + d_{CE} + d_{EB} + d_{BD} + d_{DA} = 10 + 9 + 14 + 5 + 11$$
$$= 49$$

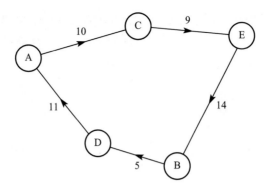

Figure 2.4

The five-town problem would appear, by inspection, to have a total of $4 \times 3 \times 2 \times 1 = 24$ possible tours but, since the distance table is symmetrical (i.e. $d_{XY} = d_{YX}$ for all pairs of towns X and Y), a tour such as $A \rightarrow D \rightarrow B \rightarrow E \rightarrow C \rightarrow A$ is equivalent to $A \rightarrow C \rightarrow E \rightarrow B \rightarrow D \rightarrow A$. It has the same total distance. Thus the number of distinct tours is only 12.

For such a small problem we could easily examine all the possible tours. The minimum distance in this case is 41. However, for a complete network of n towns it can be shown that there are $\dfrac{(n-1)!}{2}$ possible tours. To investigate all of these, even with a powerful computer capable of checking 1000 tours per second, soon becomes unreasonable. The table below illustrates the surprising amounts of time required.

Number of towns, n	$\dfrac{(n-1)!}{2}$	Computer time required to check all tours
10	181 440	3 minutes
12	$\approx 2 \times 10^7$	≈ 5.5 hours
15	$\approx 4.4 \times 10^{10}$	≈ 1.4 years
20	$\approx 6.1 \times 10^{16}$	≈ 2 million years

Clearly a method of solution based on an exhaustive evaluation of all tours is not practical. In fact, all the methods discovered to date that guarantee to find the optimum solution either take too long or need too much computer storage capacity (or both) as the size of the problem becomes larger. This is because they have factorial complexity.

The practical problem and the classical problem

The classical mathematical problem was posed by William Rowan Hamilton (1805–1865). It is to find a cycle of minimum weight (distance) visiting each vertex of a network. A cycle visiting each vertex is called a Hamilton cycle. From the definition of a cycle, vertices are visited *only once*.

> *A Hamilton cycle is a closed path (i.e. one which starts and finishes at the same vertex) which visits each vertex once and only once.*

Consider the network in figure 2.5.

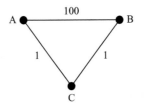

Figure 2.5

The vertices could represent towns, with A and C connected by a road and B and C connected by a road. A and B are connected only by a path. Edge weights are journey times in hours.

There is only one Hamilton cycle in this network. It is of weight 102 and can be written in six ways: $A \rightarrow B \rightarrow C \rightarrow A$, $B \rightarrow C \rightarrow A \rightarrow B$, $C \rightarrow A \rightarrow B \rightarrow C$, $A \rightarrow C \rightarrow B \rightarrow A$, $C \rightarrow B \rightarrow A \rightarrow C$, $B \rightarrow A \rightarrow C \rightarrow B$. The travelling salesperson has a better, practical, solution, however, which will take 4 hours, $A \rightarrow C \rightarrow B \rightarrow C \rightarrow A$.

You can convert the practical problem (in which vertices may be revisited) to the classical problem by replacing the network by the *complete* network of shortest distances.

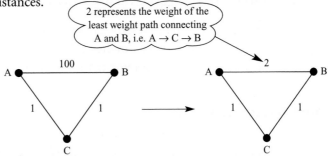

Figure 2.6

In a complete network, in which each vertex is connected to every other vertex, Hamilton cycles are certain to exist and thus there is guaranteed to be a minimum weight Hamilton cycle. The minimum weight Hamilton cycle in the network of shortest distances gives us the solution to the underlying practical problem.

A minimum weight tour in the network of shortest distances is $A \rightarrow B \rightarrow C \rightarrow A$. It is of length 4 and corresponds to $A \rightarrow C \rightarrow B \rightarrow C \rightarrow A$ in the original network.

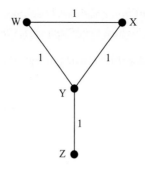

Figure 2.7

Figure 2.7 shows another example.

There is no Hamilton cycle in this network, though there are several solutions to the practical problem, for example W → X → Y → Z → Y → W, of length 5.

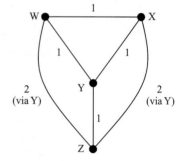

You can convert the network in figure 2.7 to a complete network of shortest distances as shown in figure 2.8.

Figure 2.8

A minimum weight Hamilton cycle in this network is W → X → Y → Z → W, of length 5. This corresponds to W → X → Y → Z → Y → W in the original problem.

You can use Floyd or Dijkstra *n* times to convert a practical problem to the classical problem. If you can then solve the classical problem you can interpret the solution to give a solution to the underlying practical problem.

Upper and lower bounds

Since it is difficult to find the optimal solution to large problems it is useful to be able to say that we know that the shortest tour lies between two distances. These are called upper and lower bounds.

A lower bound, *L*, is a distance less than or equal to the length of the shortest tour. An upper bound, *U*, is a distance greater than or equal to the length of the shortest tour. So you can write

$$L \leqslant \text{length of shortest tour} \leqslant U.$$

You will want to find a lower bound which is as large as possible and an upper bound which is as small as possible to narrow down the interval in which you know the length of the shortest tour to lie.

Lower bounds

Figure 2.9 shows a small part of a large and complete network. The edge weights have been omitted.

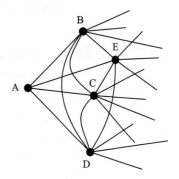

Since this is a complete network there *will* be a minimum weight Hamilton cycle and it *will* pass (once only) through A. So if you remove A together with all of A's edges from the network you will remove two edges from the Hamilton cycle. The remaining edges of the cycle will form a (very simple) tree, which connects all of the remaining vertices.

Figure 2.9

 You might think about showing this on the complete graph of which figure 2.9 is a part, but you do not know which of the edges incident on A are edges in the minimum weight Hamilton cycle, only that two of them are.

So:

the sum of the weights of the remaining edges of the cycle $\geq$ the weight of a minimum spanning tree for the remaining network

the weights of the two edges deleted from the cycle $\geq$ the weights of the two edges of least weight that were deleted.

You can use this to write an algorithm for finding a lower bound for the minimum weight Hamilton cycle in a complete network.

Step 1   Choose a vertex.

Step 2   Delete that vertex and all of its edges from the network.

Step 3   Use Kruskal's or Prim's algorithm to find the weight of a minimum connector for what is left.

Step 4   Add in the weights of the two least weight deleted edges.

Step 5   Repeat, if required, deleting a different vertex.

Step 6   Choose the largest of the lower bounds.

Note

If the arcs of the minimum connector plus the two least weight deleted arcs happen to form a tour, then that will be a minimum weight tour. However, this is unlikely except in very simple networks.

The bread delivery network from Exercise 2B is shown again in figure 2.10. The distances are all shortest distances and there is a Hamilton cycle, so the lower bound method will work.

❓ Construct a counter-example to show that the lower bound method does not always work if there is no Hamilton cycle.

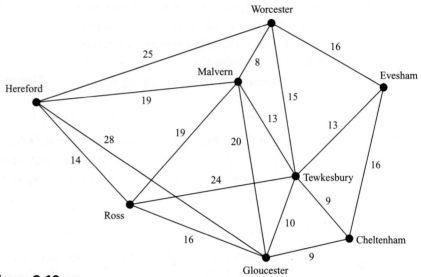

Figure 2.10

Step 1   Choose Hereford, say.

Step 2   Deleting that vertex and all of its edges from the network gives the network shown in figure 2.11.

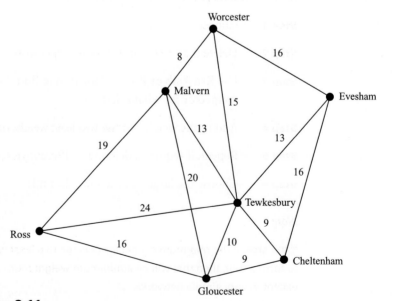

Figure 2.11

Step 3 Use Kruskal's or Prim's algorithm to find the weight of a minimum connector for what is left. The outcome is shown in figure 2.12. The minimum connector is unique in this case. Its total length is 68.

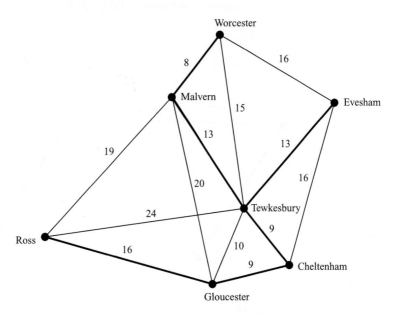

Figure 2.12

Step 4 Adding in the weights of the two deleted edges of least weight gives $68 + 14 + 19 = 101$. So a lower bound for the least weight Hamilton cycle is 101.

Step 5 Apply the algorithm again this time deleting Gloucester, say.

Step 2 Deleting that vertex and all of its edges from the network gives the network shown in figure 2.13.

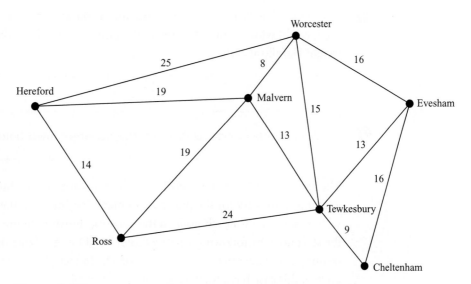

Figure 2.13

Step 3   Use Kruskal's or Prim's algorithm to find the weight of a minimum connector for what is left. There are two different minimum connectors in this case. They both have total length 76. One of the outcomes is shown in figure 2.14.

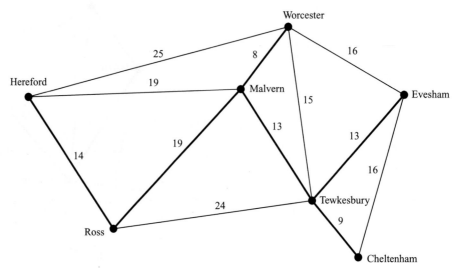

Figure 2.14

Step 4   Adding in the weights of the two deleted edges of least weight gives 76 + 9 + 10 = 95. So another lower bound for the least weight Hamilton cycle is 95.

Step 6   Choosing the largest of the lower bounds, the greatest lower bound found so far is 101.

⚠ This method does not produce a tour, unless you are *very* lucky. It produces a total weight which cannot be greater than the total weight of the minimum weight Hamilton cycle.

❓ Try deleting other towns. Is there a better, i.e. larger, lower bound?

The lower bound technique works for the classical problem, but is not guaranteed to work for the practical problem. For instance, the practical salesperson can visit each node in the network shown in figure 2.15 and return to the start node by following a path of total weight 8. Applying the lower bound technique and deleting T gives a value of 32. (In fact, the minimum Hamilton cycle in this case has a total weight of 32.)

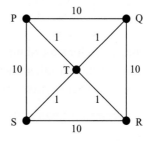

Figure 2.15

In the network shown in figure 2.16, deleting Y and its arcs would leave it disconnected, so it is not possible to find a minimum connector for the remainder.

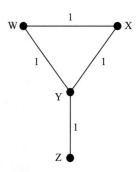

Figure 2.16

Upper bounds

It is often argued that twice the weight of a minimum connector is an upper bound for the total weight of the minimum Hamilton cycle. This is true for the practical problem, but not very useful since it tends to give a large upper bound. It is not guaranteed to give an upper bound for the classical problem.

For instance, the minimum connector for the network shown in figure 2.17 has total weight 2, so twice this is 4. The minimum Hamilton cycle, however, has total weight 102.

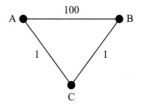

Figure 2.17

A better approach is to convert to the classical problem. The lower bound can be found using the method described on page 55. This works for the classical problem. The upper bound can be found using the method described in the next section. This works for the classical problem and generally produces low upper bounds.

The method produces a tour and the total weight of any tour is an upper bound for that of the least weight tour.

Tour building

The starting point for a tour-building algorithm is an arbitrary town. From this town you successively include other towns until a tour is achieved. A very simple scheme of this type is the *nearest neighbour rule*. It is an example of a greedy algorithm.

Step 1   Select any town as the starting point.

Step 2   From the towns not already in the sequence, find the town nearest to the last town selected. Add it to the sequence. Make an arbitrary choice in the event of a tie.

Step 3   Repeat step 2 until all towns are included in the sequence, then join the first and last towns to form a tour.

The resulting tour depends on the town chosen to start the sequence. The algorithm is a heuristic algorithm. It does not guarantee to produce an optimal solution.

The nearest neighbour algorithm is only *guaranteed* to produce a Hamilton cycle if the network is complete, that is, if each pair of vertices has a connecting arc, and if only direct links are used in steps 2 and 3.

⚠ The nearest neighbour algorithm *looks* very similar to Prim's algorithm for finding a minimum connector, but they *are* very different. Using the nearest neighbour algorithm you go next to a vertex which is nearest to the vertex you are at. Using Prim's algorithm you go next to a vertex which is nearest to your current connected set. Do not get confused between them!

Look at the bakery problem again, shown in figure 2.18.

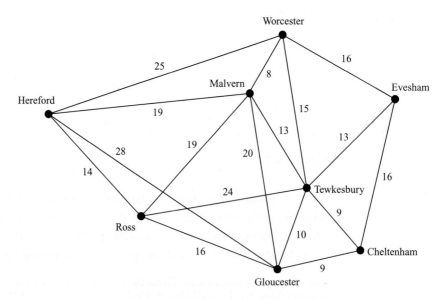

Figure 2.18

The shortest distance and route matrices are:

	C	E	G	H	M	R	T	W
C	18	16	9	37	22	25	9	24
E	16	26	23	41	24	37	13	16
G	9	23	18	28	20	16	10	25
H	37	41	28	28	19	14	32	25
M	22	24	20	19	16	19	13	8
R	25	37	16	14	19	28	24	27
T	9	13	10	32	13	24	18	15
W	24	16	25	25	8	27	15	16

	C	E	G	H	M	R	T	W
C	C	E	G	G	T	G	T	T
E	C	T	T	W	W	T	T	W
G	C	T	C	H	M	R	T	T
H	G	W	G	R	M	R	M	W
M	T	W	G	H	W	R	T	W
R	G	T	G	H	M	H	T	M
T	C	E	G	M	M	R	C	W
W	T	E	T	H	M	M	T	M

Starting at Worcester and using the distance matrix: (It helps to have a copy of the matrix, and to cross out columns as you visit the corresponding nodes.)

$$W \xrightarrow{8} M \xrightarrow{13} T \xrightarrow{9} C \xrightarrow{9} G \xrightarrow{16} R \xrightarrow{14} H \xrightarrow{41} E \xrightarrow{16} W$$

This is a total distance of 126 miles. All of the links are direct except for H → E. The route matrix shows you that this portion represents H → W → E.

So an upper bound for the least weight Hamilton cycle is 126.

Alternatively, starting at Evesham,

$$E \xrightarrow{13} T \xrightarrow{9} C \xrightarrow{9} G \xrightarrow{16} R \xrightarrow{14} H \xrightarrow{19} M \xrightarrow{8} W \xrightarrow{16} E$$

These are all direct connections. The total length is 104 miles, which is close to the lower bound found earlier, so you probably would not look any further.

Although this tour was found by starting the *algorithm* from Evesham, the *actual* tour can start from any of the nodes. Thus the 104-mile baker's tour is W → E → T → C → G → R → H → M → W, or the reverse.

Tour to tour improvement

The starting point for a tour to tour improvement algorithm is an arbitrary tour. The solution generation scheme is a rule for finding a shorter tour by making a modification to the present tour, for example by interchanging the order in which two towns are visited. The procedure terminates when application of the rule for finding a shorter tour yields no further improvement.

Tour to tour improvement methods tend to be rather tedious for manual computation since the rule for finding a shorter tour may have to be applied many times. Any change in order is as likely to increase as to decrease the tour distance, so much effort is employed in rejecting unprofitable amendments. The methods can, however, easily be programmed for computer solution and in this form can prove highly efficient. The final tour is dependent on the tour chosen as the starting point.

Two possible generation schemes are suggested below.

Scheme 1

Exchange arcs that connect towns in the tour with other arcs not in the tour whose distance is less than the arcs removed.

For example, suppose the present tour is $A \rightarrow B \rightarrow C \rightarrow D \rightarrow E \rightarrow A$ (see figure 2.19). Consider removing two arcs that do not have a town in common, such as $E \rightarrow A$ and $C \rightarrow D$.

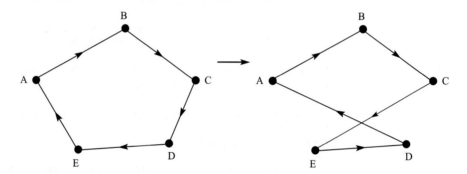

Figure 2.19

If $d_{CE} + d_{DA} - (d_{EA} + d_{CD}) < 0$ connect $D \rightarrow A$ and $C \rightarrow E$. The new tour is $A \rightarrow B \rightarrow C \rightarrow E \rightarrow D \rightarrow A$.

Scheme 2

This consists of interchanging pairs of towns in the current tour. The first town in the sequence remains fixed throughout. You consider the possible improvement brought about by swapping the town in second position with the town in each subsequent position, then make the most advantageous interchange. The process is repeated until no improvement can be made by swapping the town in the second position.

You then try swapping the town in the third position with towns in the subsequent positions, again opting for the most advantageous change. This is repeated until no further improvement can be made. You then move on to the fourth position and so on until you reach the $(n-1)$th position (where n is the number of towns). The whole process is then repeated until no more improvements are achieved by swapping pairs of towns.

1 Find upper and lower bounds for the five town problem given on page 51.

2 Investigate bounds for the following problem. A traveller from London plans to visit five towns on one of her business trips and wishes to minimise the length of her journey. The places she has to visit are Chester, Dover, Glasgow, Oxford and Plymouth. The inter-town distances are given in the table below.

	London	Chester	Dover	Glasgow	Oxford	Plymouth
London	–	182	70	399	56	214
Chester	182	–	255	229	132	267
Dover	70	255	–	472	127	287
Glasgow	399	229	472	–	356	484
Oxford	56	132	127	356	–	179
Plymouth	214	267	287	484	179	–

3 A group of tourists staying in Weston wishes to visit all the places shown on the following map. Suggest a route that will minimise their total driving distance.

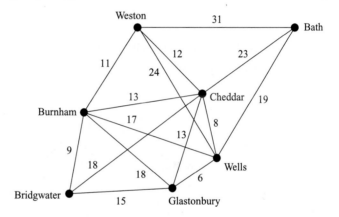

4 A depot located in Birmingham supplies goods to customers in Sheffield, Nottingham, Stoke, Shrewsbury, Hereford, Gloucester and Northampton. Plan a suitable route for the delivery lorry if it has to make deliveries in all of these towns on one trip.

The distances involved are shown in the table.

	Birmingham	Sheffield	Nottingham	Stoke	Shrewsbury	Hereford	Gloucester	Northampton
Birmingham	–	77	50	43	43	52	52	50
Sheffield	77	–	37	47	79	125	128	94
Nottingham	50	37	–	50	79	102	102	57
Stoke	43	47	50	–	34	83	89	85
Shrewsbury	43	79	79	34	–	52	75	93
Hereford	52	125	102	83	52	–	28	91
Gloucester	52	128	102	89	75	28	–	72
Northampton	50	94	57	85	93	91	72	–

5 The following problem appeared on the back of a packet of Quaker Oats some years ago. The problem is to calculate the shortest possible route, starting and finishing in St Hélier, visiting all the marked places. Although it is easy to find the answer by inspection, the problem is an interesting one to try with the algorithms. It would be a good idea to start by simplifying the network.

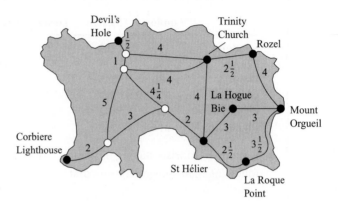

6 Find the shortest travelling salesperson tour, starting and finishing at A, for the network below.

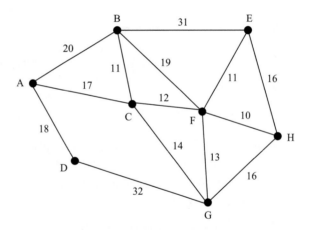

7 In a sweet-making factory, five flavours of fruit drop are made one after another on a single machine. After each flavour, the machine must be cleaned in readiness for the next flavour. The time spent cleaning depends on the two flavours as indicated in the table below.

Time in minutes	Next flavour to be made				
	Strawberry	Lemon	Orange	Lime	Raspberry
Last flavour made					
Strawberry	–	14	12	19	16
Lemon	21	–	14	10	19
Orange	19	16	–	17	20
Lime	17	9	13	–	15
Raspberry	20	15	13	19	–

The production manager wishes to find a sequence which minimises the total time spent cleaning the machine in each cycle from strawberry to strawberry, making each flavour of fruit drop once only per cycle.

(i)   By constructing an appropriate network, explain how the problem may be formulated as a travelling salesperson problem. Hence, by using the nearest unvisited city heuristic starting from strawberry, suggest a production sequence to the manager.

Someone notices that the smallest cleaning time is in changing from lime to lemon. Accordingly, he suggests that a better production sequence may be found by using the nearest unvisited city heuristic beginning from lime so that the sequence will begin with the change from lime to lemon.

(ii) Determine whether he is right that a better sequence will be found.

[JMB]

8 Apply some of the algorithms to the following famous travelling salesperson problem, known as *The Barachet Ten Town Problem*. The table of distances is given below. (The matrix is symmetrical, but only the top half is shown.)

	A	B	C	D	E	F	G	H	I	J
A	–	28	57	72	81	85	80	113	89	80
B		–	28	45	54	57	63	85	63	63
C			–	20	30	28	57	57	40	57
D				–	10	20	72	45	20	45
E					–	22	81	41	10	41
F						–	63	28	28	63
G							–	80	89	113
H								–	40	80
I									–	40
J										–

9 A depot located at town A supplies goods to customers in towns B, C, D and E. The inter-town distances are given in the table below.

	A	B	C	D	E
A	–	28	57	20	45
B	28	–	47	46	73
C	57	47	–	76	85
D	20	46	76	–	40
E	45	73	85	40	–

Usually a single vehicle will suffice for a particular delivery but today the customers' requirements are 100 units each and the vehicle available will only carry 300 units. Another similar vehicle can be hired locally, but how should the two vehicles be routed?

(Hint: introduce an artificial depot.)

The route inspection problem

Note

The route inspection problem is sometimes known as the Chinese postman problem: 'postman' because it is the problem which a postman faces, 'Chinese' because a Chinese mathematician first wrote about it.

A heavy frost is forecast for the Isle of Wight and the local council in Newport has decided to send out its two gritting lorries to treat the main roads. One will treat the left-hand and the other the right-hand side of the roads. They will follow the same route in opposite directions. What route should they take?

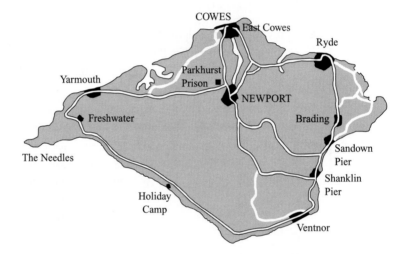

Figure 2.20

The problem faced here is to start from a node, travel along each arc and return to the starting point. If it is possible to achieve this without going over the same arc twice, the minimum distance is just the sum of the arc lengths. If it is necessary to cover some arcs twice, then you must select these in the most economic way. This type of problem is called a *route inspection problem*. Before you consider how to approach route inspection problems, you need to look at some network theory.

Networks

Node type

An *n*-node is a node where *n* arcs join. Figure 2.21 shows a 3-node.

1-nodes, 3-nodes (and so on) are referred to as *odd* nodes; 2-nodes, 4-nodes (and so on) are referred to as *even* nodes.

Figure 2.21

Numbers of nodes

In any network there is always an even number of odd nodes. This can be proved as follows.

Adding together the node type numbers of all the nodes in a network will give twice the number of arcs. This is because each arc is counted twice, once from each end. Thus the sum of node type numbers is even for any network.

You cannot have an odd number of odd nodes, because a sum containing an odd number of odd numbers would be odd. Thus any network must have an even number of odd nodes.

Traversability

A network is said to be *traversable or Eulerian* if you can draw it without removing your pen from the paper and without tracing the same arc twice.

Is the network shown in figure 2.22 traversable?

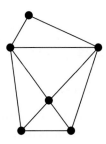

Figure 2.22

You have probably found that the network is traversable, but you must start at one odd node and end at the other. (Note that it has two odd nodes.) If you investigate drawing other networks and note their node type numbers for each node, you will find the following rule.

> For a network to be traversable it must have zero or two odd nodes and, if you are to be able to start and finish at the same node, it must have no odd nodes.

Application of network theory to route inspection problems

The implications of this result about traversability for route inspection problems are as follows.

If there are no odd nodes in the network, the network is traversable. The minimum distance is the sum of the arc distances, so the problem is trivial.

Otherwise you know there will be an even number of odd nodes, and the route inspection algorithm requires that you identify them and link them together in pairs in the most economic way. The links selected will be repeated in the final route. The effect of adding these extra arcs is to make all nodes even and hence the network becomes traversable.

Now you are ready to try the following problem. A power station located at Z provides power for seven towns located at A, B, C, D, E, F and G. After some blizzards the engineer wants to go on a tour of inspection of the pylons and power lines. These are represented by the arcs in figure 2.23. Their lengths, in kilometres, are shown. What is the shortest route the engineer can take for her inspection?

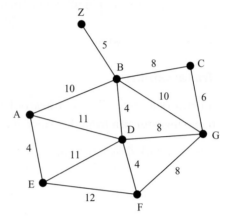

Figure 2.23

The odd nodes are Z, A, B, D, E and F as shown in figure 2.24.

Look at the distances between these nodes. If there is no direct link, you can find the shortest route between each pair by inspection. In a more complex example, or on a computer, Dijkstra's or Floyd's algorithms could be used to find the shortest paths.

Now pair up the six nodes in such a way as to give the smallest possible extra distance. This example is a simple one: it is fairly easy to see from the network diagram (without the help of the table) that the best pairing is ZB, AE and DF. In a more difficult example, a systematic analysis of the possibilities from a table would have to be undertaken.

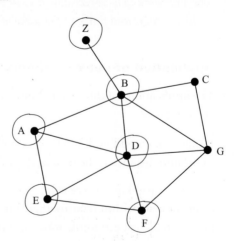

Figure 2.24

	Z	A	B	D	E	F
Z	–	15	5	9	19	13
A		–	10	11	4	15
B			–	4	14	8
D				–	11	4
E					–	12
F						–

The length of the shortest route is the sum of all the arc lengths plus the lengths of ZB, AE and DF.

Shortest route = 101 + 5 + 4 + 4
         = 114

One possible route of this length is shown by the sequence of circled numbers in figure 2.25, but there are many.

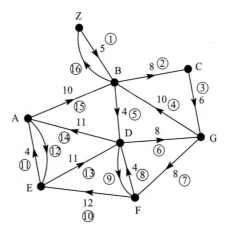

Figure 2.25

1 Solve the route inspection problem for the two networks given below.

(i)

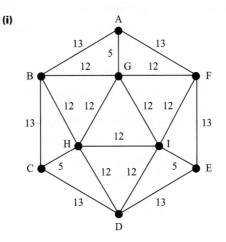

(ii)

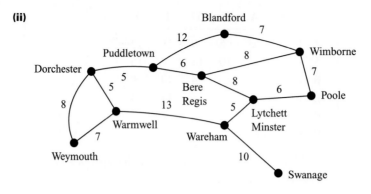

2 The airport plan below shows two runways R01 and R02 and taxi-ways to and from the apron in front of the terminal building.

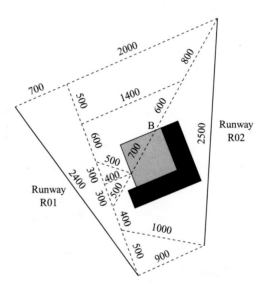

Each night when there are no flights, the runways and taxi-ways have to be inspected. The crew responsible for this job is based at B. What is the length of the shortest route for this inspection

(i) if a single drive down each runway is sufficient

(ii) if the runways must be covered twice, once in each direction?

3 Solve the route inspection problem for the following network, given in tabular form. Try to work without drawing the network but, if you are having difficulties, you will find it on page 51.

	Malvern	Worcester	Hereford	Evesham	Ross	Tewkesbury	Gloucester	Cheltenham
Malvern	–	8	19	∞	19	13	20	∞
Worcester	8	–	25	16	∞	15	∞	∞
Hereford	19	25	–	∞	14	∞	28	∞
Evesham	∞	16	∞	–	∞	13	∞	16
Ross	19	∞	14	∞	–	24	16	∞
Tewkesbury	13	15	∞	13	24	–	10	9
Gloucester	20	∞	28	∞	16	10	–	9
Cheltenham	∞	∞	∞	16	∞	9	9	–

4   A highways maintenance depot must inspect all the manhole covers within its area. The road network is given below. In order to make the inspection an engineer must leave the depot, D, drive along each of the roads in the network at least once and return to the depot. What is the minimum distance that he must drive? Give a route which enables him to drive this distance.

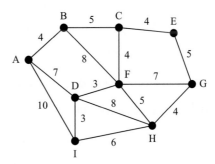

5   The island authorities of Elsea are based in St Oz and wish to inspect their road network for fallen trees after a night of strong gales. Advise on a suitable route, which should be as short as possible.

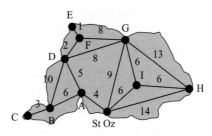

6 A policeman has to patrol the streets on the map below. Can you advise him on an efficient route?

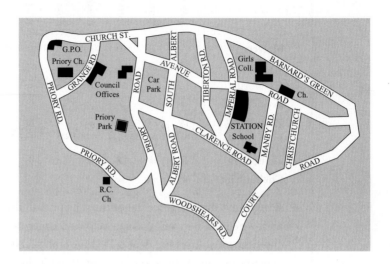

Complexity of the route inspection algorithm

The central task in the algorithm is the pairing of odd nodes.

If there are two odd nodes then there is only one pairing.

If there are four odd nodes, A, B, C and D, then there are three possible pairings:

AB│CD; AC│BD; AD│BC.

If there are six odd nodes, A, B, C, D, E and F, then there are fifteen possible pairings:

AB│CD│EF   AC│BD│EF   AD│BC│EF   AE│BC│DF   AF│BC│DE
AB│CE│DF   AC│BE│DF   AD│BE│CF   AE│BD│CF   AF│BD│CE
AB│CF│DE   AC│BF│DE   AD│BF│CE   AE│BF│CD   AF│BE│CD

With six odd nodes, the number of possible pairings is $5 \times 3 \times 1$. With eight odd nodes the number of possible pairings is $7 \times 5 \times 3 \times 1 = 105$. Generalising, with $2n$ odd nodes the number of possible pairings is $(2n-1) \times (2n-3) \times (2n-5) \times \dots \times 5 \times 3 \times 1$. You can write this rather more succinctly by multiplying and dividing as follows:

$$\frac{2n \times (2n-1) \times (2n-2) \times (2n-3) \times (2n-4) \times \dots \times 6 \times 5 \times 4 \times 3 \times 2 \times 1}{2n \quad \times \quad (2n-2) \quad \times \quad (2n-4) \times \dots \times 6 \quad \times \quad 4 \quad \times \quad 2}$$

$$= \frac{(2n)!}{2(n) \times 2(n-1) \times 2(n-2) \times \dots \times 2(3) \times 2(2) \times 2(1)}$$

$$= \frac{(2n)!}{2 \times 2 \times 2 \times \dots \times 2 \times n \times (n-1) \times \dots \times 2 \times 1}$$

$$= \frac{(2n)!}{2^n (n)!}$$

This shows that the algorithm has factorial complexity and that it is not therefore of practical use on big problems. For instance, with 40 odd nodes the number of pairings to investigate would be $\dfrac{(2 \times 20)!}{2^{20} \times (20!)} \approx 3.2 \times 10^{23}$.

For a large problem you would have to be content with a *heuristic* algorithm. This would *not* guarantee to produce the optimal solution, but would produce a good solution efficiently.

INVESTIGATION

Develop and test a heuristic algorithm to solve large-scale route inspection problems.

1 The Chinese postman (route inspection) problem is to find the shortest route which traverses every edge in a network at least once. An algorithm for this involves pairing the vertices of odd degree in the network. Thus, if a network has four odd vertices, A, B, C and D, then one pairing is A with B and C with D.

   (i)   How many ways are there of pairing four odd vertices?
   (ii) How many ways are there of pairing six odd vertices?
   (iii) How many ways are there of pairing eight odd vertices?
   (iv) How many ways are there of pairing twenty odd vertices?

   [AEB, adapted]

2 The graph is a representation of a system of roads. The lengths of the roads are shown in metres.

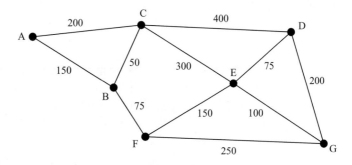

   (i)   List the odd vertices in the graph.
   (ii) Explain why the graph is not Eulerian (i.e. not traversable).
   (iii) By considering ways of pairing the odd vertices, find the shortest route, starting and finishing at A, and traversing each road at least once. State the length of your route.

   [AEB]

3 The following matrix represents the distances, in kilometres, between the six nodes of a network.

	A	B	C	D	E	F
A	5	11	21	16	9	10
B	11	0.5	10	31	7	23
C	21	10	57	14	12	20
D	16	31	14	11	11	9
E	9	7	12	11	18	12
F	10	23	20	9	12	43

(i) Delete vertex A from the network and use Prim's algorithm to find a minimum connector for the remaining five vertices. Start from B and indicate the order in which you include vertices.
State which arcs are in your minimum connector and give its total length.

(ii) By considering the connections between A and the other five vertices, construct a lower bound for the optimal solution of the travelling salesperson problem for the network.

(iii) Explain why your answer to part (ii) is a lower bound for the length of the optimal solution of the travelling salesperson problem for the network.

(iv) Use the nearest neighbour algorithm to find a solution to the travelling salesperson problem which starts and finishes at A. Calculate its length and compare it with the lower bound which you found in part (ii).

[Oxford]

4 The map shows a number of roads in a housing estate. Road intersections are labelled with capital letters and the distances in metres between intersections are shown. The total length of all of the roads in the estate is 2300 m.

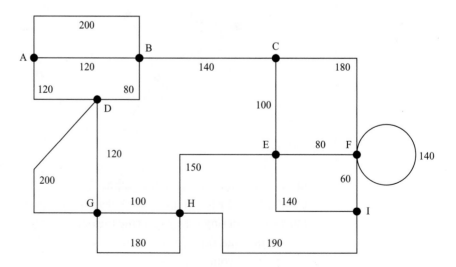

(i) A newspaper deliverer has to walk along each road at least once, starting and ending at A. The shortest route to achieve this is required.

    (a) List those intersections which are of odd order and explain their significance to the problem.

    (b) By investigating all possible pairings of odd intersections, find the minimum distance which the newspaper deliverer has to walk. (You are not required to apply a shortest distance algorithm to solve this. You are required to show the distance computations which lead to you choosing a particular pairing of odd intersections.)

    (c) For each intersection other than A, give the number of times that the newspaper deliverer must pass through that intersection whilst following the shortest route.

(ii) The newspaper deliverer only calls at a proportion of houses. The postwoman has to call at most houses and, since the roads are too wide to cross continually back and forth, she finds it necessary to walk along each road twice, once along each side. She requires a route to achieve this in the minimum distance.

    (a) Describe how to produce a network to model this problem.

    (b) Without drawing such a network, say why it will be traversable and calculate the minimum distance which the postwoman will have to walk.

(iii) The streetcleaner needs to drive his vehicle along both sides of each road. He has to drive on the correct side of the road at all times. He too would like a shortest route. Explain how the streetcleaner's problem differs from the postwoman's and say how the network would have to be modified to model this.

[MEI]

5 (i) When moored alongside a harbour wall a sailing boat has to be secured by four ropes. The ropes are attached to the points A, B, C and D shown in the diagram.

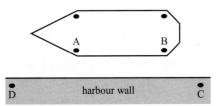

The point A has to be attached to D, and also to C. The point B has to be attached to C and also to D.

(a) Regard the mooring ropes as defining the arcs of a network and the points A, B, C and D as being the nodes. Give the orders of the nodes.

(b) Explain, in terms of the orders of the nodes, why the network is traversable. Indicate the implications of this if only one long length of rope is available instead of four shorter lengths.

(ii) When mooring alongside another boat ropes are attached both to the other boat and also to the harbour wall as shown.

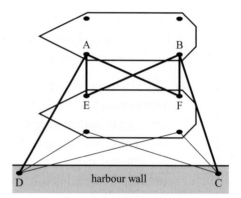

Regard the ropes marked with bold lines in the diagram, together with the points A, B, C, D, E and F, as a network.

(a) Give the orders of the nodes and explain why the existence of odd nodes means that the network is not traversable.

(b) What practical consequence does the non-traversability of the network have on the use of a single length of rope to moor the boat?

(c) The solution to the route inspection problem for the network gives a way of mooring the boat with a single length of rope.
Solve the route inspection problem for the network, starting and finishing at the same node, and give the corresponding sequence of nodes.

(d) Suppose that in part (ii)(c), the route is allowed to start at C and finish elsewhere. Give two solutions, both starting at C but finishing at different nodes, in which no more than two edges are repeated.

(iii) When using a single length of rope to moor the boat against the other boat the rope may be run, if required, directly from A to B, or directly from C to D. If the rope is to run directly from A to B at some point, and is to be as short as possible, give the two possible starting points, and the corresponding ending points. Give a corresponding order of nodes.

[MEI]

6 Two friends, holidaying in the town marked J on the map, hire a car to explore the island.

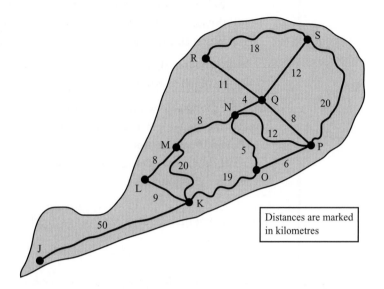

(i) (a) The network below represents the island's road system. Use Dijkstra's algorithm to find the shortest distances from town J to each of the other marked towns. Show your working on a copy of the boxes.

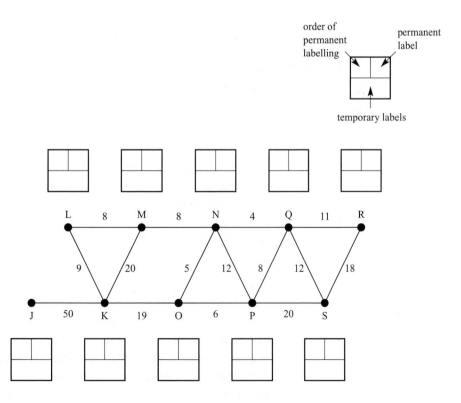

(b) Give the shortest distance from J to each of the other towns.

(ii) One of the friends wants to visit every town on the island starting and finishing at J. She uses the following algorithm.

> Go to the nearest town not yet visited (even if this means passing through a town that has already been visited).
> When all towns have been visited return to J by the shortest route.

Write down in order the towns that she passes through and give the total length of her journey.

(iii) The other friend would like to travel along every road.
Use a suitable algorithm to find the minimum distance he must travel, showing your method clearly.
Write down, in order, the towns that he passes through and give the total length of his journey.

[MEI]

7 The following matrix gives the costs of flight tickets for direct flights between six connected cities.

From \ To	A	B	C	D	E	F
A	–	45	60	58	90	145
B	45	–	67	25	83	100
C	60	70	–	50	70	320
D	50	25	50	–	35	210
E	100	80	70	35	–	72
F	145	110	300	175	80	–

In this question a tour is a journey from a city, visiting each other city once and only once, and returning to the starting city.

(i) Use the nearest neighbour algorithm to find a tour, starting and finishing at A, with a low associated cost.
Show that the algorithm has not produced the minimum cost tour.

(ii) How many different tours which start and finish at A are there altogether?

(iii) Suppose that, in addition to the cost of tickets, airport taxes must be paid on leaving an airport according to the following table.

A	B	C	D	E	F
20	30	20	40	10	20

Thus the flight from A to B will cost 45 for the ticket and 20 tax, a total of 65.
Produce a matrix showing total costs, i.e. the total of fares and taxes.

(iv) Give an example where the cheapest route from one city to another differs when taxes are taken into account from that when taxes are not taken into account. Show your two routes and give their costs.

(v) Do airport taxes have an effect on the problem of finding the cheapest tour starting and finishing at A? Why?

[MEI]

8 The following matrix gives fares for direct bus journeys between towns, P, Q, R, S, T, U and V, in a tourist area. Blanks indicate no direct service.

Fares in pence

	P	Q	R	S	T	U	V
P		57		35			70
Q	57		53	61		160	
R		53			49		
S	35	61			30		44
T			49	30		48	52
U		160			48		40
V	70			44	52	40	

(i) Draw a network to represent this information.

(ii) Using Dijkstra's method, find the minimum cost route from P to U. Describe in full detail all of the steps of the method, specifying the order in which you assign values to vertices.

(iii) A holidaymaker sets out from V to travel over all of the routes in the area, returning to V at the end. Find the minimum cost for the journey and the route that should be taken.

Explain the significance of towns P and U in your solution.

(iv) A 'summertime special' service is to be provided linking R to S at a cost of 35p. What is the new minimum cost and the best route for the holidaymaker in part (iii)? (In this part the necessary minimum connectors may be found by inspection.)

[Oxford]

9 The weights on the network represent distances.

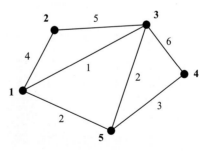

(i) Give a walk of minimum length which traverses every arc at least once, and which returns to the start. Give the length of your walk.

(ii) The initial tables and the results of iterations 1, 2, 4 and 5 when Floyd's algorithm is applied to the network are shown overleaf.

	1	2	3	4	5
1	∞	4	1	∞	2
2	4	∞	5	∞	∞
3	1	5	∞	6	2
4	∞	∞	6	∞	3
5	2	∞	2	3	∞

	1	2	3	4	5
1	1	2	3	4	5
2	1	2	3	4	5
3	1	2	3	4	5
4	1	2	3	4	5
5	1	2	3	4	5

	1	2	3	4	5
1	∞	4	1	∞	2
2	4	8	5	∞	6
3	1	5	2	6	2
4	∞	∞	6	∞	3
5	2	6	2	3	4

	1	2	3	4	5
1	1	2	3	4	5
2	1	1	3	4	1
3	1	2	1	4	5
4	1	2	3	4	5
5	1	1	3	4	1

	1	2	3	4	5
1	8	4	1	∞	2
2	4	8	5	∞	6
3	1	5	2	6	2
4	∞	∞	6	∞	3
5	2	6	2	3	4

	1	2	3	4	5
1	2	2	3	4	5
2	1	1	3	4	1
3	1	2	1	4	5
4	1	2	3	4	5
5	1	1	3	4	1

	1	2	3	4	5
1					
2					
3					
4					
5					

	1	2	3	4	5
1					
2					
3					
4					
5					

	1	2	3	4	5
1	2	4	1	7	2
2	4	8	5	11	6
3	1	5	2	6	2
4	7	11	6	12	3
5	2	6	2	3	4

	1	2	3	4	5
1	3	2	3	3	5
2	1	1	3	3	1
3	1	2	1	4	5
4	3	3	3	3	5
5	1	1	3	4	1

	1	2	3	4	5
1	2	4	1	5	2
2	4	8	5	9	6
3	1	5	2	5	2
4	5	9	5	6	3
5	2	6	2	3	4

	1	2	3	4	5
1	3	2	3	5	5
2	1	1	3	1	1
3	1	2	1	5	5
4	5	5	5	5	5
5	1	1	3	4	1

(a) Copy and complete the two tables for iteration 3.

(b) Use the final route table to give the shortest route from vertex 4 to vertex 2.

(c) Use the final distance table to draw a complete network with weights representing the shortest distances between vertices.

(iii) Using the complete network of shortest distances, find a lower bound for the solution to the travelling salesperson problem by deleting vertex 1 and its arcs, and by finding the length of a minimum connector for the remainder. (You may find the minimum connector by inspection.)

(iv) Use the nearest neighbour algorithm, starting at vertex 1, to produce a Hamilton cycle in the complete network. Give the length of your cycle.

(v) Interpret your Hamilton cycle in part (iv) in terms of the original network.

[MEI]

10 White, yellow, blue, green and red dyes are to be used separately in a dyeing vat (a container in which materials are dyed). Each colour is to be used once during each day. The vat has to be cleaned between colours, and the cost of this depends on which colour was previously used, and on which colour is going to be used. For example, if the blue dye was previously used, and yellow dye is going to be used next, the cost is 5. These costs, in suitable units, are shown in the table. The vat must be returned to its original colour at the end of the day.

		To				
		W	Y	B	G	R
	W	–	0	2	1	2
	Y	4	–	4	3	4
From	B	8	5	–	1	2
	G	7	3	1	–	3
	R	7	4	3	3	–

(i) Explain why this problem is similar to the travelling salesperson problem.

(ii) Use the nearest neighbour algorithm five times, starting from each colour in turn, to find a low-cost sequence of colours.

(iii) Give a colour sequence of cost 12, starting and ending with white.

(iv) Because the network is directed, the technique of deleting a vertex and finding a minimum connector for the remainder to produce a lower bound will not work. Explain why not.

[MEI]

11 Direct transport links between five cities are shown in the diagram. The weights on the arcs are the times in hours for moving along those links.

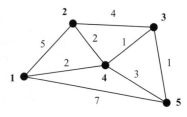

(i) Set up initial time and route matrices for an application of Floyd's algorithm to find the complete network of shortest times between cities.

The tables below show completed time and route matrices at the end of the second iteration of Floyd's algorithm, and incomplete third iteration matrices.

	1	2	3	4	5
1	10	5	9	2	7
2	5	10	4	2	12
3	9	4	8	1	1
4	2	2	1	4	3
5	7	12	1	3	14

	1	2	3	4	5
1	2	2	2	4	5
2	1	1	3	4	1
3	2	2	2	4	5
4	1	2	3	1	5
5	1	1	3	4	1

	1	2	3	4	5
1			9		
2			4		
3	9	4	8	1	1
4			1		
5			1		

	1	2	3	4	5
1			2		
2			3		
3	2	2	2	4	5
4			3		
5			3		

(ii) Copy and complete the matrices for the third iteration.

The final time and route matrices after completion of the fifth iteration of Floyd's algorithm are shown below.

	1	2	3	4	5
1	4	4	3	2	4
2	4	4	3	2	4
3	3	3	2	1	1
4	2	2	1	2	2
5	4	4	1	2	2

	1	2	3	4	5
1	4	4	4	4	4
2	4	4	4	4	4
3	4	4	4	4	5
4	1	2	3	3	3
5	3	3	3	3	3

(iii) Use the final matrices to find the fastest route and shortest time from city 5 to city 2.

(iv) By temporarily deleting city 4 and its arcs, find a lower bound for the minimum duration Hamilton cycle in the complete network of shortest times, as given in the final matrices. (You do not need to show the use of a minimum connector algorithm.)

(v) Use the nearest neighbour algorithm, starting at city 1, to find a Hamilton cycle in the complete network of shortest times, as given in the final matrices. Give the time for your cycle.

(vi) Use your answer to part (v) to construct a fast route for an individual wishing to leave city 4, visit each of the other cities, and return to city 4.

(vii) Find the shortest time needed to traverse every link in the original network, starting and finishing at the same vertex.

[MEI]

12 The map shows a small island. The labelled points are beaches. Tracks are marked, with distances shown in km. The total length of the tracks is 32 km.

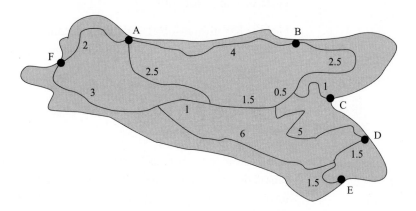

Floyd's algorithm is applied to a six-node network extracted from this map, giving the following final matrices.

Distance matrix

	A	B	C	D	E	F
A	4	4	5.5	9	11	2
B	4	7	3.5	8	11	6
C	5.5	3.5	7	6.5	9.5	7
D	9	8	6.5	6	3	10.5
E	11	11	9.5	3	6	10.5
F	2	6	7	10.5	10.5	4

Route matrix

	A	B	C	D	E	F
A	F	B	C	D	E	F
B	A	C	C	D	D	A
C	A	B	B	D	D	F
D	A	B	C	E	E	F
E	A	D	D	D	D	F
F	A	A	C	D	E	A

(i) Explain how to use the matrices to find the shortest distance and route from E to B. Give the route and the distance.

Joan and Keith spend a few days on the island. Joan wants to visit every beach.

(ii) Use the nearest neighbour algorithm on the shortest distance network, given by the final distance matrix, to find an efficient order for Joan to visit all of the beaches, starting and finishing at beach A. (Note that this may give a route which involves re-visiting a vertex in the original network.)

Give the distance travelled, and use the route matrix to give the route.

(iii) Give a shorter route starting and finishing at A.

Keith would like to walk along every track on the island.

(iv) Explain why the six-node network would not be a suitable model for finding a good route for Keith. Give the number of nodes on an appropriate network. Which of the nodes A, B, C, D, E and F would not be needed?

(v) Obtain an efficient route for Keith, starting and finishing at A, and give its length.

[MEI]

1 All shortest paths in a network can be found either by repeatedly applying Dijkstra's algorithm, or by applying Floyd's algorithm.

2 Floyd's algorithm involves updating a distance matrix and a route matrix at each iteration. The update involves checking whether it is better to go via the current vertex.

3 The travelling salesperson problem (TSP) is to find a minimum weight Hamilton cycle in a complete network, that is, one in which each pair of vertices has a connecting arc.

4 A Hamilton cycle visits each vertex once and once only, returning to the start. The practical problem allows vertices to be revisited, and need not involve a complete network. The practical problem can be converted to the classical by constructing the complete network of least weight paths.

5 Solving the TSP by complete enumeration is not practical. It has factorial complexity.

6 The total weight of any Hamilton cycle is an upper bound for the solution to the TSP. A lower bound can be constructed by deleting a vertex and its arcs, finding the weight of a minimum connector of the remainder, and adding back in the least two deleted weights.

7 The nearest neighbour algorithm – go next to the nearest vertex not yet visited and return to the start vertex when all have been visited – is a heuristic algorithm. It is only guaranteed to produce a Hamilton cycle if the network is complete, but it usually produces good approximations.

8 The route inspection (Chinese postman) problem is to find the least weight cycle which traverses every arc of a network. If there are no odd nodes then this can be done using each arc just once and the network is traversable (Eulerian).

9 The route inspection algorithm involves investigating all different pairings of odd nodes. Within a pairing the least weight path connecting each pair of nodes is found. The pairing having the least total of these least weights is chosen, and the corresponding least weight paths are traversed twice in the solution.

10 The route inspection algorithm is a complete enumeration algorithm, and has factorial complexity. It is not, therefore, practical for large problems. These need heuristic techniques.

3 Decision analysis

A wise man makes his own decisions, an ignorant one follows public opinion.

Chinese proverb

An oil prospector is drilling for oil in a certain region. So far she has drilled ten wells and found oil in three of them. If oil is found at a site she makes a net profit of £2 million, but if no oil is found she makes a net loss of £1 million.

The prospector is given the chance to explore a new site. Based on the available evidence would you advise her to drill a well or not?

A geological survey team has just moved into the area, offering to carry out a survey of the new site for £150 000. If the result of the survey is promising then they estimate that there is a 60% chance of finding oil, but if the survey produces an unpromising outcome then there is only a 20% chance of finding oil at the site. The probability of a promising survey is calculated to be 25%.

Now what do you think the oil prospector should do: should she drill a well, commission a survey or forego the chance to explore the site?

Decisions have to be made on many situations like the one above where a chance element arises. In such situations there are often stages at which we must make a decision and some stages at which the outcome depends purely on chance. For example, the oil prospector has to decide whether to commission a survey or not (this is a decision stage), but the result of the survey is affected by chance. Once the survey result is known then she must decide whether or not to drill, but if she drills then chance determines whether oil is found.

In order to advise the oil prospector you need to develop some appropriate methods, and you can do this by studying some simpler situations first.

EXPERIMENTS

Two simple games of chance are described below. Play each game several times with a partner, one of you acting as banker (but do not use real money!).

When you have played each game try to answer the following questions:

(i) In whose favour is this game: the player's or the banker's?
(ii) How much does the player win or lose per game on average?

Game 1

You pay the banker £5 to flip two coins. For two heads you win £5 (and get your £5 stake back), for a head and a tail you win £2 (and get your stake back), for two tails your £5 stake is forfeit.

Game 2

You throw a die. If a 5 or 6 shows the banker pays you £18; for any other number you lose £6. However, in the latter case you may if you wish throw the die again. This time if you throw a 6 you win £36 but otherwise you lose a further £6.

Analysis of the games

Game 1

You can tabulate the profit to the player for each outcome as shown in the table.

Result	HH	HT or TH	TT
Probability	$\frac{1}{4}$	$\frac{1}{2}$	$\frac{1}{4}$
Profit (£)	+5	+2	−5

The player's average profit per game (or his expected profit) is therefore

$$\left(\tfrac{1}{4} \times £5\right) + \left(\tfrac{1}{2} \times £2\right) + \left(\tfrac{1}{4} \times -£5\right) = £1$$

So the game is in the player's favour and he can expect to gain £1 per game on average in the long run. The name given to the average gain in the long run is the *expected monetary value* (EMV for short). Hence, in this game, the player's EMV is £1 per game.

An alternative method of analysing this game is to use a tree diagram as in figure 3.1.

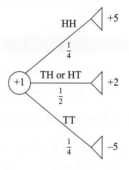

Figure 3.1

A triangle symbol is used to represent the outcomes and the EMV of +1 is written in a circle or *chance node*. You could extend the diagram as shown in figure 3.2.

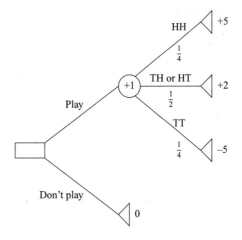

Figure 3.2

Figure 3.2 shows that the player can decide whether to play or not. There are now three types of node in the diagram:

- *triangles* to denote end pay-offs to the player

- *rectangles* to denote stages at which the player makes a decision – *decision nodes*

- *circles* to denote stages where chance determines the outcome – *chance nodes*.

This kind of diagram is called a *decision tree*. To use it to analyse the problem you work from the right-hand side by labelling the pay-off boxes, then you calculate the EMV of +1 as you did above and insert this in the chance node.

You now move back to the rectangular decision node and see that 'Play' has an EMV of +1, and 'Don't play' an EMV of 0. On this basis you would decide to choose 'Play' and a double line is put through 'Don't play' to show that this is an inferior option. This is shown in figure 3.3. You again, of course, conclude that the game is in the player's favour with an EMV of £1 per game.

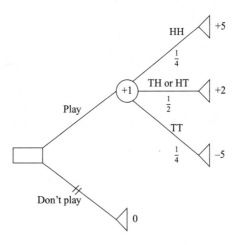

Figure 3.3

Game 2

Using the techniques described above you can now try to analyse game 2. The decision tree is shown in figure 3.4. Calculate the EMVs and work out whether the player should agree to play this game before you read on!

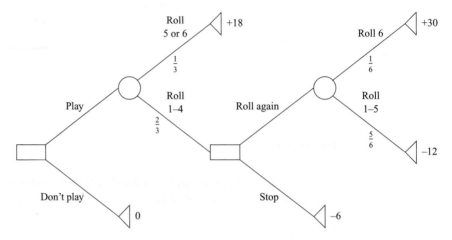

Figure 3.4

To work through this diagram, first find the EMV for the chance node on the right. This is

$$\left(\frac{1}{6} \times 30\right) + \left(\frac{5}{6} \times -12\right) = -5$$

Label this chance node with '–5'.

Moving back to the decision node in the middle of the diagram, you see that 'Roll again' has an EMV of –5 and 'Stop' has an EMV of –6. The double line is therefore placed across the 'Stop' option to denote that this is the inferior option.

To complete the diagram the EMV for the chance node on the left of the diagram must be found. This is

$$(\tfrac{1}{3} \times 18) + (\tfrac{2}{3} \times -5) = \tfrac{8}{3}$$
$$= 2\tfrac{2}{3}$$

The final diagram is shown in figure 3.5.

The analysis suggests the following strategy:

'Play the game, and if you roll a 1, 2, 3 or 4 on the first throw then roll again.'

With this strategy the player's EMV is £$2\frac{2}{3}$ per game.

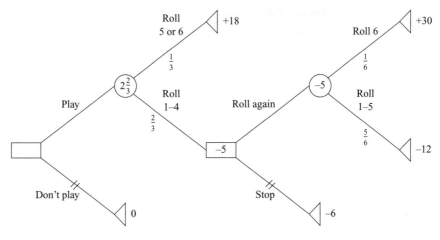

Figure 3.5

The procedure developed above is usually referred to as the *EMV algorithm*. At each stage you take the decision which maximises your EMV.

Utility

There is an important assumption implicit in using EMV to measure the worth of alternative courses of action, namely that £1 is worth the same to everybody. You might argue that it is obviously worth the same since it will buy the same. But others might argue that a rich person would not regard £1 as being worth very much. To model this decision analysts often use the concept of *utility* to measure the relative worth of alternative actions. The analyst has to investigate the decision-maker's *utility function*. This gives the value of an extra £1 as a function of the wealth held by the decision-maker. It is always an increasing function of the wealth, but the slope is usually decreasing.

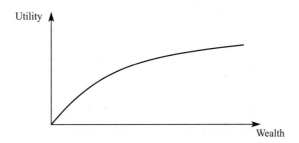

Figure 3.6 *A typical utility function*

Determining a decision-maker's utility function is an exercise in psychology rather than in mathematics but, having determined it, the function can be applied and the course of action which maximises utility can be chosen.

For example, consider the possibility of playing a game in which you win £10 with probability 0.05, or lose £1 with probability 0.95. Would you choose to play?

One possible analysis, using the EMV criterion, is represented by the tree diagram in figure 3.7. This shows the EMV of playing a game, but it does not analyse the decision (although the decision is obvious – do not play!).

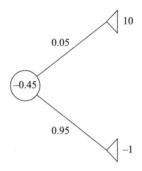

Figure 3.7

The decision analysis might be represented by the tree diagram in figure 3.8, in which it is assumed that you have £100 in your pocket. This says exactly the same thing: either keep your £100 in your pocket or play, in which case the expected value will drop by £0.45 to £99.55.

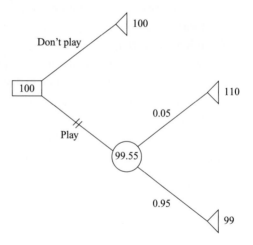

Figure 3.8

If, however, you wish to consider expected utilities, then since your utility function will be non-linear, you must use the decision analysis approach. Using utilities the analysis is not independent of your wealth. For instance, suppose that your utility function is given by $utility = \sqrt[4]{wealth^3}$. Then the analysis is shown in the tree diagram in figure 3.9 (note that $= \sqrt[4]{100^3} \approx 31.62$, etc.). In this case the decision is the same: do not play. It is almost always the case for this particular utility function, that no matter what the wealth, prize or probability of winning, the decision using the EMV criterion will be the same as the decision using the utility criterion. A more complex utility function might explain why people buy lottery tickets!

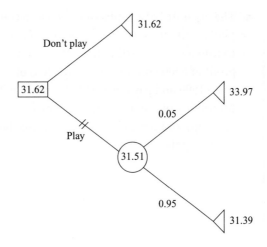

Figure 3.9

Questions 1, 2 and 3 concern expectation or EMV only, questions 4 to 9 involve decision trees.

1 (i) In a simple game the player chooses a playing card from a normal pack of 52. For an ace he wins £20, for a king or queen £15, and for a jack he pays the banker £50. If any other card is chosen he neither wins nor loses. What is the EMV of the game?

(ii) A roulette wheel has 37 numbers into which the ball can drop. Eighteen of these are coloured red, eighteen are coloured black and the number 0 is not coloured. A player bets £1 on red. If the ball lands in a red number then he wins £1 and gets his £1 stake back, but for any other number his £1 stake is forfeit. Find the EMV of the game.

2 (i) You play a game in which you will win £15 if you throw a total score of 11 or 12 with two dice; otherwise you will lose £2. What is the EMV of the game?

(ii) What would your winnings need to be in part (i) instead of £15 to make your EMV zero?

(iii) You have just £2. Your utility function is given by $utility = \sqrt[4]{wealth^3}$. What would your winnings need to be to tempt you to risk your last £2 on a game?

3 (i) Three normal unbiased dice are thrown. Explain why

$$P(three\ 6s) = \tfrac{1}{216}$$

and find $P(no\ 6s)$, $P(one\ 6)$ and $P(two\ 6s)$.

(ii) In the game 'Crown and Anchor' a player pays £1 to throw three normal unbiased dice. If three 6s turn up he wins £3 and gets his £1 stake back, if two 6s turn up he wins £2 and gets his £1 stake back, if one 6 turns up he wins £1 and gets his £1 stake back. If no 6s turn up his £1 stake is forfeit. Find the EMV of the game.

4 The organiser of a tennis tournament has to decide whether to take out pluvius insurance (that is insurance against rain on the day of the tournament). She estimates that on a fine day the tournament will make a profit of £5000 but on a wet day a loss of £10 000. The pluvius insurance will cost £1500 and pay out £6000 if it rains on the day of the tournament. The probability of rain is estimated to be $\frac{1}{5}$.

(i) Copy and complete the decision tree below to work out the organiser's best policy.

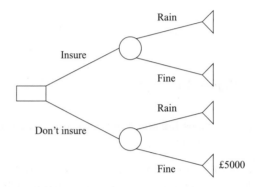

(ii) The day before the tournament a weather forecaster now estimates the probability of rain to be $\frac{1}{3}$. If the organiser had used this probability would she have made the same decision?

5 In the game of 'Sprogget' the player rolls a normal die. If it lands on a 1 she wins £1, if it lands on a 2 she wins £2 and so on. However, if she rolls a 6 she may roll the die again. On this second roll if a 6 is obtained then the player's £6 from the first round is forfeit; otherwise she wins a further £1 for a 1, £2 for a 2 and so on as in the first round.

Construct a decision tree for this game. Find the player's optimal EMV per game and state the strategy that would obtain this EMV.

6 A company has just developed a new product and now a decision must be taken as to whether to launch the product without delay, to run a market test or to abandon the product. If the product is launched without delay then market research suggests that a strong market will occur with probability 0.3 and a weak market with probability 0.7. A strong market will bring a profit of £200 000, a weak market a profit of £50 000. To run a market test will cost £5000 which will have to be deducted from profits, and this will give favourable or unfavourable indications. The probability of a favourable indication is calculated to be $\frac{1}{6}$. If the result of the market test is favourable then the probability of a strong market is now 80% but if the market test is unfavourable the probability of a weak market is 80%.

The product may be abandoned at any stage, either now or after the market test, and the technology developed may be sold to a rival company for a net profit of £90 000.

(i) Draw and label a decision tree to describe the above situation.

(ii) Use the decision tree to say what decision should be taken and find the EMV associated with this plan.

(iii) What is the maximum amount (instead of £5000) that the company should be prepared to pay to run the market test?

7 Perform a decision tree analysis on the oil prospector's dilemma at the beginning of this chapter. Hence advise the oil prospector on her best course of action.

8 At a later date the oil prospector moves to a new region. Wells are now classified as 'soaking', 'wet' or 'dry'. Soaking wells bring a profit of £2.5 million, wet wells a profit of £0.5 million and dry wells a loss of £1 million. Of the twenty wells drilled so far in this region, three have been soaking, six wet and the remaining ones dry.

(i) The prospector is given the chance to drill a new well. Based on current evidence should she accept or refuse the invitation to drill?

A geologist offers to carry out a survey of the new site. The cost of the survey is £200 000. The survey will indicate either 'promising' or 'unpromising' conditions, the probability of each of these outcomes being calculated at 0.5. The table below relates the results of the survey to the probable state of the well.

	Probability that well is soaking	Probability that well is wet	Probability that well is dry
Promising	0.2	0.4	0.4
Unpromising	0.1	0.2	0.7

(ii) Draw a decision tree to represent the situation and hence find the oil prospector's best plan.

(iii) What is the *maximum* amount that the oil prospector should be willing to pay for the survey?

9 You are fortunate enough to hold the winning ticket in a raffle in which the first prize is £100 000. As you go to claim your prize you are offered the following deal: you may either take your £100 000 prize or forego this for a second bet. In this second bet you have a 0.1 chance of winning £1 500 000 and a 0.9 chance of winning nothing.

(i) Draw a decision tree to represent the above situation and use the EMV algorithm to determine your optimal decision.

(ii) Do you think you would really follow the decision in part(i)? What if your utility function is given by $utility = \sqrt{wealth}$ and you are currently broke?

A game is played as follows. At each turn the player rolls two dice. If either die shows a 6 the player's score is zero and his turn is ended. If no 6s turn up then his score is taken to be the sum of the numbers showing, however, in this case the player may elect to roll the dice again. On this second roll the same rules apply again. If a 6 shows his score for the turn is zero and his turn is ended, otherwise the score on the dice is added to the score from the first roll. In the latter case the player may then elect to roll the dice again with the same rules, and so on. Hence the player's turn continues until either he rolls a 6 (with either die), in which case his score is zero, or he elects to stop, in which case his score is the sum of the numbers rolled so far.

Investigate the strategy the player should adopt to maximise his expected score at each turn.

1 The organiser of a tennis tournament has to decide whether to take out pluvius insurance (that is insurance against lost ticket sales due to rain on the day of the tournament). She estimates that on a fine day the tournament will make a profit of £3000, but on a wet day a loss of £7000. The pluvius insurance will cost £1000 and will pay out £8000 if it rains on the day of the tournament. The probability of rain is estimated to be 0.2.

(i) Construct a decision tree for this situation and label it with the returns, the probabilities and the EMVs (expected monetary values).
Advise her, with reasons, on whether or not she should take out the insurance.

Madame Tarot, the local fortune teller, has a reputation for success in weather forecasting. In the past if she has said that it would rain then there has been a probability of 0.5 of rain and, if she has said that it would not rain, then there has been a probability of 0.1 of rain. A quarter of her forecasts have been for rain and three quarters for no rain.

(ii) Extend your decision tree in order to analyse whether or not the organiser should pay for a consultation with Madame Tarot. Show the probabilities and EMVs on your decision tree.
(iii) How much is it worth paying for a consultation with Madame Tarot?

[MEI]

2 A government has to decide how to dispose of nuclear waste material. One possibility is to build a long term store immediately. This will cost £2500 million.

Alternatively the material can be put into temporary storage for 50 years, after which time the long term store will be built. The total cost of this alternative depends on the general level of growth in the economy, and is shown in the table.

Growth of economy	Total cost
Low growth	£6100 million
Medium growth	£2400 million
High growth	£1760 million

It is estimated that the probability of low growth is 30%, the probability of medium growth is 50%, and the probability of high growth is 20%.

(i) Construct a decision tree for this situation, marking on outcomes and probabilities, and use it to find which is the best decision for the government to take. Give the cost advantage of that decision compared to the alternative.

In reviewing the decision the government's economic spokesman expresses the opinion that low growth will not occur. He estimates that the probabilities of medium growth and high growth are each 50%.

(ii) Find the best decision, and the cost advantage of it, under these assumptions. (You are not required to draw another decision tree.)

In the debate leading to a decision the opposition party calls for an independent enquiry to be set up. The government has to decide whether or not to agree to do this. If an enquiry is set up then a decision will have to be made whether or not to accept its recommendations.

(iii) Draw a modified decision tree to show all of the alternatives:

- whether or not the government sets up an enquiry
- whether such an enquiry recommends building now or storing first
- whether the government decides to build now or to store first
- the economic conditions (assuming that high, medium and low growth are all possible).

You are not required to show outcomes or probabilities, but you are required to show the correct structure of the decision tree. Only those nodes representing decisions that are within the control of the government should be shown as decision nodes.

[MEI]

3 The management of a hotel chain want to build a new hotel on an unspoilt beach site. It has been suggested that the company should consult an environmental protection society for advice on developing environmentally friendly features, and secure their support for the plans. In the past, 60% of development projects submitted to the society have succeeded in securing the society's support.

If the society is not consulted there is a high probability (estimated at 0.60) of disruption from environmental action groups. With the society's support for the project the probability of disruption is much lower at 0.25. However, if the society is consulted and withholds its support, the probability of disruption is 0.90.

The company has two decisions to make: whether or not to consult the society and whether or not to go ahead with the project.

In the past, disruptive action has added 50% to the costs of construction projects. The company estimates that the cost of the development (without interference) will be 70 billion pesetas. Should it be decided, at any stage, to cancel the project, it is estimated that it would cost 100 billion pesetas in charges and to develop an alternative site.

(i) (a) Construct a decision tree for the company.

    (b) Find the value of consulting the society, i.e. the difference between the expected cost of consulting and the expected cost of not consulting.

(ii) An environmental action group realises that it can increase the worth of consulting the environmental society by increasing the probability p of disruption on projects which are going ahead without consultation.

    (a) What would be the value of consultation to the company if p were to be increased from 0.6 to 0.8?

    (b) Find the maximum level to which the action group can raise the value of consultation and give the minimum value of p which will achieve this.

[MEI]

4 A family doctor is a fundholder, which means that she is provided with National Health Service money for her to use. This enables her to provide treatments and tests for her patients, as long as she has not used all of her NHS money for the year. She is therefore concerned *not* to spend her NHS money if she can avoid it.

She has a patient who has tested positive for a particular disease. This indicates that there is an 80% chance that the patient has the disease and a 20% chance that he has not. The doctor has two decisions to take:

(1) She can ask for a further test which costs £50. Past experience has shown that $\frac{4}{7}$ of those sent for the further test show positive, and $\frac{3}{7}$ show negative.

Records show that 95% of patients who test positive under this further test have the disease. However, 60% of those testing negative also have it.

(2) Whether or not she has the results of a further test, and whether such results are negative or positive, she can either send the patient for treatment or wait to see if the disease develops.

Sending the patient for treatment costs £500. If she waits and the disease does not develop it will cost nothing. If she waits and the disease does develop then the later treatment will be more expensive at £800.

(i) On a copy of the decision tree on the opposite page, mark at the right-hand end of each route through the tree the total costs incurred as a consequence of going along that route.

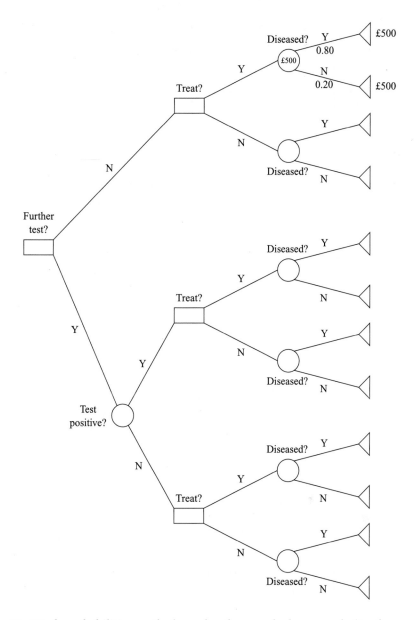

(ii) Mark probabilities on the branches from each chance node (i.e. from each node marked with a circle).

(iii) Compute the expected cost at each node.

(iv) Advise the doctor on her cheapest course of action, and say what the expected cost will be.

[MEI]

5 A decision has to be made regarding a project. It can be allowed to proceed (A), or it can be cancelled (~A). Outcomes, payoffs and probabilities are summarised in the decision tree overleaf.

(i) If the EMV of proceeding is equal to the EMV of cancelling, show that $p = 0.375$.

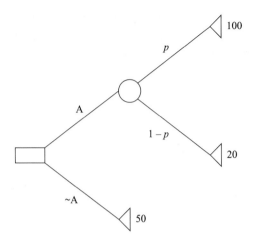

(ii) In fact, $p = 0.25$. Advice can be sought on whether or not to proceed. If advice is sought then there is a probability of 0.2 that the advice will be to proceed, in which case the resultant probabilities will be more favourable. If the advice is to cancel, then the probabilities are less favourable. The values are summarised in the decision tree below.

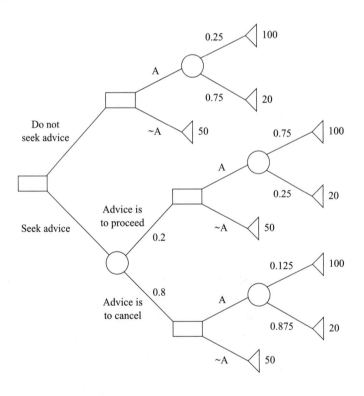

(a) Copy and complete the second decision tree, showing the EMVs.

(b) What is the value of the advice?

(c) To what value would the cancellation payoff of 50 have to increase to make it not worth seeking advice?

[MEI, adapted]

6 A fairground game offers three outcomes, £10, £1 or £0. A player has just been offered £1. She can accept or 'go again', in which case the offer of £1 will be withdrawn and she will be offered £10, £1 or £0 with probabilities p, q and $1 - p - q$ respectively.

(i) Draw a decision tree for this game.

(ii) If q is 0.5 what value must p exceed to tempt her to go again?

In a similar game the player has again been offered £1, but gets two chances to go again. If she elects not to take the £1 and is subsequently offered £10, she will accept it. However, if the subsequent offer is £0 she will go again for the second and last time. If the offer is £1 then she must decide whether to accept it this time or to go again for the last time.

(iii) Draw a decision tree for this game.

(iv) If $p = 0.02$ and $q = 0.5$, compute the EMV at each node and give the player's best strategy.

[MEI]

7 Despina can invest her savings of £1000 in bonds or in equities. Bonds are generally reckoned to be safer than equities. Equities have potentially higher returns (measured in % per annum), but may suffer losses (again measured in % per annum).

From past history the probability of equity returns being +15% is 0.6, but the probability of equity returns being −10% is 0.4. For bonds the probability of a 5% increase is 0.8 and the probability of a 3% increase is 0.2.

(i) What will be the value of Despina's savings after one year under each of the four possible scenarios, i.e.

- if she invests in equities and they increase in value by 15%

- if she invests in equities and they decrease in value by 10%

- if she invests in bonds and they increase in value by 5%

- if she invests in bonds and they increase in value by 3%.

(ii) Draw a decision tree for Despina. Use the EMV criterion to advise her where to invest her money.

Despina's utility function for her savings is given by

$$utility = (monetary\ value)^p,$$

where $p < 1$.

(iii) Using expected utility, show that the value of p which will make Despina indifferent between investing in equities and investing in bonds is 0.5 (correct to 1 d.p.).

[MEI]

8 One of three similar types of new car, A, B or C, is to be purchased. The decision is to be made on the basis of annual service and repair costs. For each car a warranty can be purchased which insures against unexpected costs. Otherwise a chance can be taken on whether the particular car purchased turns out to be reliable or unreliable. The costs and probabilities are shown in the table below.

	Annual costs (£)			Probabilities	
	With warranty	**Reliable**	**Unreliable**	**Reliable**	**Unreliable**
A	1000	750	2100	$\frac{2}{3}$	$\frac{1}{3}$
B	1100	800	2000	$\frac{3}{4}$	$\frac{1}{4}$
C	1100	810	1800	$\frac{5}{6}$	$\frac{1}{6}$

(i) Copy and complete the decision tree below and give the best decision, together with its EMV.

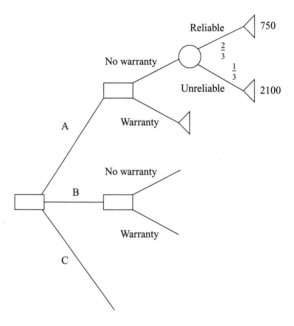

An alternative is to buy a cheaper second-hand car. Annual service and repair costs of second-hand cars are higher, and warranties are more expensive. A free independent inspection of one car can be arranged. Approval by the inspector gives a good indication of the car being reliable. If the report is not favourable then a warranty will be purchased, fixing costs at £1150 per year. The relevant probabilities are summarised on the decision tree on the opposite page.

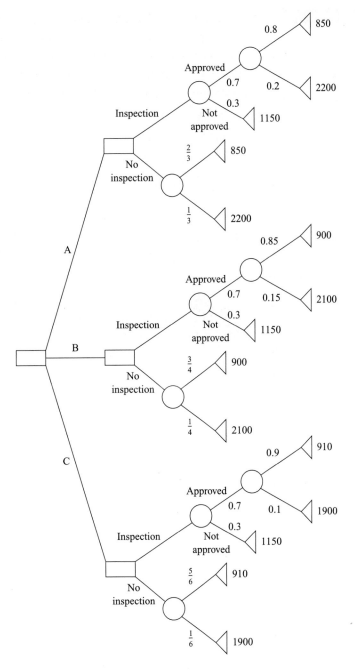

(ii) Copy the second decision tree. Complete the EMV calculations and give the best course of action and its EMV.

(iii) For each type of car give the value of having an inspection.

(iv) The cost of a warranty increases. To what value would the fixed cost of £1150 per year have to rise to change the decision in part (ii)?

[MEI]

KEY POINTS

1 When the outcomes of events are uncertain, and where probabilities can be assigned to those outcomes, the EMVs (expected monetary values) of courses of action can be computed. Decisions can be chosen which maximise the EMV or which maximise the utility of the outcomes.

2 In analysing a decision-making process a decision tree is constructed. This has three types of node:

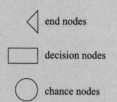

end nodes

decision nodes

chance nodes

3 When the tree is finished it is valued by working from right to left. EMVs (or utilities) are computed at chance nodes. Best courses of action are selected at decision nodes by placing a 'stop' on inferior choices.

4 Logic

... contrariwise, if it was so it might be; and if it were so, it would be: but as it isn't, it ain't.
That's logic.

Tweedledee in Through the Looking-glass, Lewis Carroll

In an ancient puzzle, a life or death decision is to be made, whether to turn left or right at a junction. Two sentinels stand at the junction, one of whom always lies and the other of whom always tells the truth. They both know the way to salvation, but only one question is allowed.

❓ What question would you ask?

Now imagine that there are three sentinels, one who always tells the truth, one who always lies and one who randomly lies or tells the truth.

❓ What question or questions would you need to ask? (Don't spend long on this – it's a trick question!)

In logic you attempt to model the way in which we reason and argue. In these first steps the discrete set is the set of propositions, simple statements which may either be true or false. This area of mathematics is known as *propositional calculus*.

Here are two statements.

> *p: The weather is fine.*
> *q: You have an umbrella with you.*

These are simple statements which, as you read this, will probably either be true or false – *probably* because, for instance, it might be a changeable day. A *proposition* has no such ambiguity. It is either true or false.

In studying logic you should not be overly concerned with which truth value (i.e. *true* or *false*) is allocated to a specific proposition, but rather with the consequences. You might wish, for instance to arrange things so that whenever *p* is true then *q* is false, and whenever *p* is false, *q* is true.

Propositional calculus

A calculus is defined to be 'a method of computation'. (It is derived from the latin word for a *stone* since stones were used as calculation aids in the same way that an abacus is still used in some countries today.) It might seem odd to think of computing propositions, but much of mathematics is concerned with sets of elements and with rules for combining those elements. You will be used to using sets of numbers and combining them using operations such as addition and multiplication. You can do similar things with propositions.

To model the process of thinking and reasoning, start with a set consisting of statements which may either be true (denoted by '1') or false (denoted by '0') and then consider ways of combining such propositions to make new propositions.

Here are some further, weather-related propositions:

m: It is warm today.
n: It is wet today.

Clearly the truth or falsity of these statements depend on the day on which you are reading this, together with what you regard as warm and what you think of as being wet. But you need not concern yourself with the allocation of a 1 or a 0 to each proposition. Instead you will be concerned with the truth or falsity of propositions such as *It is neither warm nor wet today* under *all* possible allocations of 0 and 1 to those basic propositions.

The proposition *It is neither warm nor wet today* is called a compound proposition. Compound propositions are constructed by using logical operators, some of which are called *connectives*.

These operators are defined by showing the truth values of the compound proposition under all possible circumstances, that is for all possible combinations of truth values of the underlying propositions. This listing is called a *truth table*.

Truth tables

Negation: NOT, ~

p	~p
0	1
1	0

If *p* is *The weather is fine*, then *~p* can be read as *The weather is not fine* or *It is not the case that the weather is fine.*

Conjunction: AND, ∧

p	q	p ∧ q
0	0	0
0	1	0
1	0	0
1	1	1

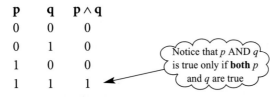

Notice that *p* AND *q* is true only if **both** *p* and *q* are true

If *q* is *You have an umbrella with you*, then *p ∧ q* is *The weather is fine and you have an umbrella with you.*

From the table you can see that the full set of possibilities (four in this case) can be listed in an ordered way by using binary numbers: 00, 01, 10, 11.

Note

∧ (AND) in logic corresponds to ∩ (intersection) in set theory.

Disjunction: OR, ∨

p	q	p ∨ q
0	0	0
0	1	1
1	0	1
1	1	1

p ∨ q means *Either the weather is fine or you have an umbrella with you.*

Note

∨ (OR) in logic corresponds to ∪ (union) in set theory.

⚠ This models the natural use of *or*, where the compound proposition is true if either *or both* of the parts are true. It is possible to define and use the exclusive OR, which is true if either of the constituent propositions are true, but not both. The exclusive OR is not used in this book.

With the set of real numbers × and + are sufficient for our needs. (Subtraction can be achieved by adding the negative, and division by multiplying by the reciprocal.) In the same way you could make do, in propositional calculus, with ~, ∧ and ∨. (You will see why in the section on proof.) However, it is convenient to introduce three further connectives.

Equivalence: ⇔

p	q	p ⇔ q
0	0	1
0	1	0
1	0	0
1	1	1

$p \Leftrightarrow q$ means *You have an umbrella with you if and only if the weather is fine* or *The weather is fine is equivalent to you having an umbrella with you.*

You need to be careful not to let linguistic details distract you from the essential structure of a compound statement, which is the same in both these cases.

Note

$p \Leftrightarrow q$ is often stated as '*p* if and only if *q*' and abbreviated as *p* IFF *q*.

Implication: ⇒

p	q	p ⇒ q
0	0	1
0	1	1
1	0	0
1	1	1

$p \Rightarrow q$ means *If the weather is fine then you have an umbrella with you* or *The weather is only fine if you have an umbrella with you.*

The definition of implication causes much debate. The final two rows are generally accepted but the first two often create difficulties. Consider another example:

> *a* is the proposition *You work hard at this module.*
> *b* is the proposition *You pass this module.*
> $a \Rightarrow b$ is the proposition *If you work hard at this module then you will pass this module.*

What happens if you do not work hard and still pass? Does this make the compound proposition false?

You have to allow $a \Rightarrow b$ to be true when *a* is false to model the fact that, if *a* is false, it is possible for *b* to either be true or false. $a \Rightarrow b$ says nothing about what happens if you do not work hard. You might fail. You might be lucky and pass anyway.

Sufficiency: $\Leftarrow$

p	q	$p \Leftarrow q$
0	0	1
0	1	0
1	0	1
1	1	1

$p \Leftarrow q$ means *The weather is fine if you have an umbrella with you* or *It is sufficient for you to have an umbrella with you for the weather to be fine.*

$p \Leftarrow q$ is true when the weather is fine and you do not have an umbrella with you because it is not stating that it is *necessary* for you to be carrying an umbrella for it to be fine.

If you find it difficult to see that the first and second of each of the two statements are the same, try thinking about what each line of the truth table means.

EXERCISE 4A The statements p, q, m and n are defined as follows.

> p: The weather is fine.
> q: You have an umbrella with you.
> m: It is warm today.
> n: It is wet today.

Use logical operators, together with p, q, m and n as defined above, to write out the following propositions.

1 If it is warm today then it is fine.
2 If it is not wet today then it is fine.
3 If you have an umbrella with you then it is not wet.
4 You have an umbrella with you if and only if it is wet and not warm.
5 You only have an umbrella when it is not fine.

Write out the following propositions in English. (Again, with p, q, m and n as defined above.)

6 $\sim q \Rightarrow p$
7 $\sim p \Rightarrow n$
8 $(m \wedge \sim n) \Rightarrow p$
9 $q \Leftarrow n$
10 $(q \wedge n) \Leftrightarrow \sim p$

Proof

Now that you have a model for propositions and for compound propositions, you can start to model what is meant by *proof*. You might think, for example, that the proposition $\sim m \wedge \sim n$ (*It is cold and dry today*) is equivalent to the proposition $\sim(m \vee n)$ (*It is neither warm nor wet*). You can *prove* this by showing that the

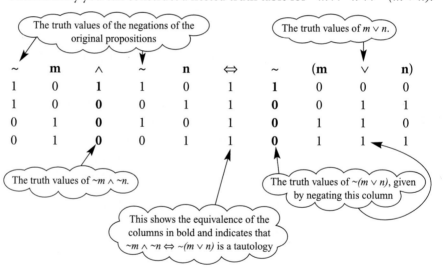

propositions have the same truth values whatever the truth values of *m* and *n*.
Alternatively you can show that $\sim m \wedge \sim n \Leftrightarrow \sim (m \vee n)$ is true, no matter what the
truth values of *m* and *n* happen to be.

Note

A proposition which is always true, no matter what the truth values of its constituent
parts happen to be, is called a *tautology*.

Here is the truth table for $\sim m \wedge \sim n$:

m	n	~m	~n	~m ∧ ~n
0	0	1	1	1
0	1	1	0	0
1	0	0	1	0
1	1	0	0	0

Here is the truth table for $\sim (m \vee n)$:

m	n	m ∨ n	~(m ∨ n)
0	0	0	1
0	1	1	0
1	0	1	0
1	1	1	0

These two columns are the same

This proves that the compound propositions *~m ∧ ~n and ~(m ∨ n)* are
equivalent, no matter what the individual propositions *m* and *n* represent.

Alternatively you can construct a nested truth table for $\sim m \wedge \sim n \Leftrightarrow \sim (m \vee n)$:

The truth values of the negations of the
original propositions

The truth values of *m ∨ n*.

~	m	∧	~	n	⇔	~	(m	∨	n)
1	0	**1**	1	0	1	**1**	0	0	0
1	0	**0**	0	1	1	**0**	0	1	1
0	1	**0**	1	0	1	**0**	1	1	0
0	1	**0**	0	1	1	**0**	1	1	1

The truth values of *~m ∧ ~n*.

The truth values of *~(m ∨ n)*, given
by negating this column

This shows the equivalence of the
columns in bold and indicates that
~m ∧ ~n ⇔ ~(m ∨ n) is a tautology

EXAMPLE 4.1

Prove that $\sim p \vee q$ is equivalent to $p \Rightarrow q$.
Note that $\sim p \vee q$ means $(\sim p) \vee q$. You would need the brackets for $\sim (p \vee q)$.

SOLUTION

p	q	~p	~p ∨ q
0	0	1	1
0	1	1	1
1	0	0	0
1	1	0	1

This also shows that you could get
by without using ⇒, but it is easier
to use it than to avoid it

This gives the same outcomes as the table defining $p \Rightarrow q$, showing that $\sim p \vee q$
is equivalent.

Laying this out in a nested table:

~	p	∨	q	⇔	p	⇒	q
1	0	**1**	0	**1**	0	**1**	0
1	0	**1**	1	**1**	0	**1**	1
0	1	**0**	0	**1**	1	**0**	0
0	1	**1**	1	**1**	1	**1**	1

The column of 1s under the ⇔ shows that the expression agrees under all circumstances and is, therefore, a tautology.

EXAMPLE 4.2

Prove $(a \wedge b) \vee c \Leftrightarrow (a \vee c) \wedge (b \vee c)$.

SOLUTION

Eight rows are needed, since there are three constituent propositions and $2^3 = 8$.

Looking separately at the left and right-hand sides:

a	b	c	a∧b	(a∧b)∨c	a∨c	b∨c	(a∨c)∧(b∨c)
0	0	0	0	0	0	0	0
0	0	1	0	1	1	1	1
0	1	0	0	0	0	1	0
0	1	1	0	1	1	1	1
1	0	0	0	0	1	0	0
1	0	1	0	1	1	1	1
1	1	0	1	1	1	1	1
1	1	1	1	1	1	1	1

Alternatively, in nested form:

(a	∧	b)	∨	c	⇔	(a	∨	c)	∧	(b	∨	c)
0	0	0	0	0	1	0	0	0	0	0	0	0
0	0	0	1	1	1	0	1	1	1	0	1	1
0	0	1	0	0	1	0	0	0	0	1	1	0
0	0	1	1	1	1	0	1	1	1	1	1	1
1	0	0	0	0	1	1	1	0	0	0	0	0
1	0	0	1	1	1	1	1	1	1	0	1	1
1	1	1	1	0	1	1	1	0	1	1	1	0
1	1	1	1	1	1	1	1	1	1	1	1	1

EXERCISE 4B

1 Prove $a \vee \sim a$ is a tautology.

2 Prove $\sim(a \vee b) \Leftrightarrow \sim a \wedge \sim b$.

3 Prove $(a \Rightarrow b) \Leftrightarrow (\sim b \Rightarrow \sim a)$.

Note

$\sim b \Rightarrow \sim a$ is called the *contrapositive* of $a \Rightarrow b$.

4 Prove $[a \wedge (a \Rightarrow b)] \Rightarrow b$.

5 Prove $a \wedge (b \vee c) \Leftrightarrow (a \wedge b) \vee (a \wedge c)$.

6 Draw up a truth table for $[p \vee (\sim p \vee q)] \vee (\sim p \wedge \sim q)$.

7 In his book *Symbolic logic* Lewis Carroll gave arguments such as:

> *No misers are unselfish.*
>
> *None but misers save eggshells.*
> _____
> *No unselfish people save eggshells.*

The question is whether or not the statement below the line can be deduced from those above.

> Let *m* be the proposition *He is a miser.*
>
> Let *u* be the proposition *He is unselfish.*
>
> Let *e* be the proposition *He saves eggshells.*

Then you can interpret the argument as

$$[(u \Rightarrow \sim m) \wedge (e \Rightarrow m)] \Rightarrow (e \Rightarrow \sim u)$$

Draw up a truth table for this. Is the argument valid?

8 In question 7 you could have argued in this way

$$(e \Rightarrow m) \Leftrightarrow (\sim m \Rightarrow \sim e) \text{ by the result of question 3.}$$

So you have $u \Rightarrow \sim m$
and $\sim m \Rightarrow \sim e$
giving $u \Rightarrow \sim e$
but $(u \Rightarrow \sim e) \Leftrightarrow (e \Rightarrow \sim u)$ by question 3, giving the result.

(i) Use a similar argument to prove that *the bowler has not bowled a bouncer*, given that *if the bowler bowls a bouncer then the batsman will be injured*, and that *the batsman is fit and healthy*.

(ii) What, if anything, would you have been able to deduce had the batsman been injured?

9 Write down in words, the contrapositives of the following statements.
(Reminder: $\sim b \Rightarrow \sim a$ is the contrapositive of $a \Rightarrow b$.)
(i) If a fair coin is tossed then the probability of a head is 0.5.
(ii) x and y are integers. If x is even, then xy will be even.
(iii) If triangles are congruent, then they are similar.
(iv) If you have read *Treasure Island* then you will remember the character Long John Silver.

10 Replace the following propositions by letters, and write out the compound propositions using letters and logical operators. Use truth tables to check their validity.

(i) Alice cannot have her cake and eat it.

(ii) If $x^2 = 9$ then either $x = 3$ or $x = -3$. But $x \neq 3$ so $x^2 \neq 9$.

(iii) Given that x is an integer, to prove that if x^2 is even then x is even, it is enough to prove that if x is odd then x^2 will be odd.

11 Use truth tables to check the validity of the following. In each case make up an argument to fit the implication.

(i) $[(p \vee q) \wedge \sim q] \Rightarrow p$

(ii) $[(p \Rightarrow q) \wedge p] \Rightarrow q$

(iii) $[(p \Rightarrow q) \wedge \sim q] \Rightarrow \sim p$

12 Is the statement *If $\theta = 60°$, then $\cos\theta = 0.5$* equivalent to the statement *If $\cos\theta = 0.5$, then $\theta = 60°$*? Either prove that the statements are equivalent or find a counter-example.

13 Which of the following pairs are equivalent?

(i) (a) He runs and he jumps.

(b) It is not true that he neither runs nor jumps.

(ii) (a) It is not true that he is both a philosopher and practical.

(b) He is either not a philosopher or he is not practical.

(iii) (a) It is not the case that either he is not guilty or he is not telling the truth.

(b) He is guilty and lying.

14 A student argues that $0 = 1$, so $-1 = 0$ (subtracting 1 from both sides), so $-1 = 1$ (since both $= 0$), so $1 = 1$ (squaring both sides). Explain why this does not prove that $0 = 1$.

Historical note

Lewis Carroll was the pen-name for Charles Dodgson, an Oxford mathematician who lived from 1832 to 1890. He is most famous as the author of *Alice in Wonderland* but he also wrote several mathematics books, and he was particularly interested in logic. He was an early expert in photography, specialising in children's portraits. One of his subjects was Alice Liddell, daughter of the Dean of Christ Church College, Oxford, after whom his famous fictitious character was named. Lewis Carroll also invented the word *chortle*, a mixture of chuckle and snort.

Boolean algebra

You can *prove* results by using truth tables, but you cannot *produce* results in this way. However, just as you can use the rules of the algebra of number to manipulate numerical expressions (e.g. by opening brackets), so you can use the rules of the algebra of logic to manipulate logical expressions. George Boole, wrote in 1847 of such an algebra, now called Boolean algebra.

Historical note

George Boole grew up in Lincoln in the early 1800s. He became professor of mathematics at Queens College, Cork where he was regarded as a very good teacher. It was there that he did the work for which he is best remembered, representing logic through algebraic symbols.

At the age of 49 Boole was caught in a rain storm on the way to work and gave his lectures drenched to the skin. When he got home he went to bed feeling feverish. His wife believed that the cure for any illness had to resemble the cause and so threw buckets of cold water over the bed. He did not survive this treatment.

A set, together with operations for combining members of the set to produce other members of the set, is known as an *algebra*. You know about the rules associated with the algebra of numbers, rules such as

$$a \times (b + c) = (a \times b) + (a \times c)$$
(the distributive law of multiplication over addition)

and

$$a + (b + c) = (a + b) + c$$
(the associativity of addition).

Knowing the rules of the algebra of number gives you the ability to deduce results and solve problems. For instance, you use them to solve a simple equation such as $x + 1 = 7$ as follows.

1 Add -1 to both sides. (-1 is known as the *additive inverse* of 1. The existence of additive inverses is guaranteed by a rule which is similar to the *complement rules* for Boolean algebra given in the next section.)

2 This gives $(x + 1) + -1 = 7 + -1$, which can be written as $x + (1 + -1) = 6$. (This is an application of the *associativity* of addition, which is very similar to the associative laws of Boolean algebra given in the next section.)

3 The left-hand side now simplifies to $x + 0$, where 0 is the *additive identity*, that is it has the property that $x + 0 = x$, so you end up with the result that $x = 6$. (Again, you will see in the next section that Boolean algebra has its own identity rules.)

That may seem like rather a fuss, but perhaps only because you are so familiar with the rules of number.

Knowing the rules of Boolean algebra will enable you to solve some logical problems, in just the same way that knowing the rules of number enables you to solve some numerical problems. The rules of Boolean algebra are similar to the rules for the algebra of the real numbers, but not the same.

The rules of Boolean algebra

Identity rules

$$p \wedge 1 = p \quad p \vee 1 = 1$$
$$p \vee 0 = p \quad p \wedge 0 = 0$$

These rules state what happens when a proposition is combined with a tautology (truth value 1) or with a proposition with truth value 0. For instance $p \vee 0 = p$ says that the truth value of p or a false statement is the same as the truth value of p.

Associative rules

$$(p \vee q) \vee r = p \vee (q \vee r)$$
$$(p \wedge q) \wedge r = p \wedge (q \wedge r)$$

Commutative rules

$$p \vee q = q \vee p$$
$$p \wedge q = q \wedge p$$

The associative and commutative rules correspond to the ways in which addition and multiplication operate in the real numbers. Think of $\vee$ as + and $\wedge$ as ×.

Distributive rules

$$p \wedge (q \vee r) = (p \wedge q) \vee (p \wedge r)$$
$$p \vee (q \wedge r) = (p \vee q) \wedge (p \vee r)$$

In the real numbers multiplication distributes over addition, but addition does not distribute over multiplication:

$$2 \times (3 + 4) = (2 \times 3) + (2 \times 4) \text{ but}$$
$$2 + (3 \times 4) \neq (2 + 3) \times (2 + 4).$$

In Boolean algebra both distribution rules hold.

De Morgan's rules

$$\sim(p \vee q) = \sim p \wedge \sim q$$
$$\sim(p \wedge q) = \sim p \vee \sim q$$

You have already proved the first of these on page 109.

Double negation

$$\sim(\sim p) = p$$

Complement rules

$$p \vee \sim p = 1$$
$$p \wedge \sim p = 0$$

Absorption rules

$$p \wedge p = p \qquad\qquad p \wedge (p \vee q) = p$$
$$p \vee p = p \qquad\qquad p \vee (p \wedge q) = p$$

EXAMPLE 4.3

Use Boolean algebra to prove

$$(a \wedge b) \vee (\sim a \wedge \sim b) = (a \vee \sim b) \wedge (\sim a \vee b).$$

(This relates to wiring a lighting circuit with two switches. You will learn about circuits later in this chapter.)

SOLUTION

$$
\begin{aligned}
(a \wedge b) \vee (\sim a \wedge \sim b) &= [(a \wedge b) \vee \sim a] \wedge [(a \wedge b) \vee \sim b] && \text{(2nd distributive rule)} \\
&= [\sim a \vee (a \wedge b)] \wedge [\sim b \vee (a \wedge b)] && \text{(commutativity)} \\
&= (\sim a \vee a) \wedge (\sim a \vee b) \wedge (\sim b \vee a) \wedge (\sim b \vee b) && \\
& && \text{(2nd distributive rule)} \\
&= 1 \wedge (\sim a \vee b) \wedge (\sim b \vee a) \wedge 1 && \text{(complement rules)} \\
&= (\sim a \vee b) \wedge (\sim b \vee a) && \text{(identity rules)} \\
&= (a \vee \sim b) \wedge (\sim a \vee b) && \text{(commutativity rules)}
\end{aligned}
$$

EXERCISE 4C

1 Simplify the expression $a \wedge b \wedge (a \vee c) \wedge [b \vee (c \wedge a) \vee d]$.

2 Use the rules of Boolean algebra to prove the following.
  (i) $(a \wedge b \wedge c) \vee (\sim a \wedge b \wedge c) = b \wedge c$
  (ii) $\sim[(p \wedge q) \vee \sim p] = \sim q \wedge p$

3 Remembering that $(a \Rightarrow b) \Leftrightarrow (\sim a \vee b)$, use Boolean algebra to prove the following.
  (i) $[(p \Rightarrow q) \Rightarrow p] \Rightarrow p$ ← Showing that the expression equals 1 proves it is true
  (ii) $[p \Rightarrow (q \Rightarrow r)] \Leftrightarrow [(p \wedge q) \Rightarrow r]$
  (iii) $[(a \vee b) \wedge (a \Rightarrow c) \wedge (b \Rightarrow c)] \Rightarrow c$

4 Use Boolean algebra to prove the following.

(i) $(a \wedge b) \vee \sim a \vee \sim b = 1$

(ii) $(a \vee b) \wedge \sim a \wedge \sim b = 0$

(iii) $a \vee \sim[(\sim b \vee a) \wedge b] = 1$

(iv) $a \wedge \sim[(\sim b \wedge a) \vee b] = 0$

(v) $(a \vee b) \wedge (\sim a \vee \sim b) = (a \wedge \sim b) \vee (\sim a \wedge b)$

(vi) $a \wedge (\sim a \vee b) = a \wedge b$

(vii) $a \vee (\sim a \wedge b) = a \vee b$

(viii) $(a \vee b \vee c) \wedge (a \vee b) = a \vee b$

(ix) $(a \wedge b \wedge c) \vee (a \wedge b) = a \wedge b$

(x) $(a \vee b) \wedge (a \vee \sim b) = a$

(xi) $(a \wedge b) \vee (a \wedge \sim b) = a$

(xii) $a \vee [a \wedge (b \vee 1)] = a$

(xiii) $a \wedge [a \vee (b \wedge 1)] = a$

Switching and combinatorial circuits

You have seen that all that you need in propositional calculus are the operations $\sim$, $\wedge$ and $\vee$. These are the operations involved in switching and so they underpin all of digital technology.

Imagine that you are building a small electrical circuit consisting of a battery, a bulb and a switch as shown in figure 4.1.

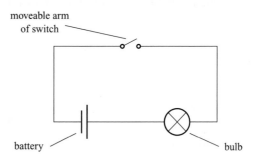

moveable arm
of switch

battery   bulb

Figure 4.1

If the switch is on (moveable arm down) the light comes on. If the arm is up the light is off. Identify the switch with a proposition, a, the bulb being on if the proposition is true and off if it is false. Figure 4.2 (overleaf) shows how you would wire switches to represent a, $\sim a$, $a \wedge b$ and $a \vee b$. Note that neither the battery nor the bulb are shown.

> *Wiring switches in series gives conjunction; wiring them in parallel gives disjunction.*

Normally you would not bother to show the moveable arm.

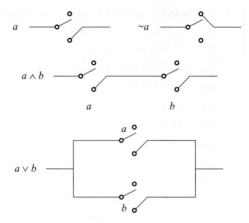

Figure 4.2

Figure 4.3 shows a switching circuit to model $a \Rightarrow b$.

Figure 4.3

Ganging switches

Switches can be *ganged* together to operate in unison. Look at figure 4.4.

Figure 4.4

This shows how c can be achieved in the upper circuit whilst $\sim c$ is achieved in the lower circuit. Having accepted that this *can* be achieved the mechanics is not of great interest, and you would draw this as shown in figure 4.5.

Figure 4.5

Logic gates

In practice mechanical switches with moving arms, whilst easy to visualise, are slow and prone to wear. Instead, switching is achieved by electronic components such as semi-conductors. These are generally grouped to produce components known as *logic gates*.

The symbols for these are shown in figure 4.6, together with an indication of their inputs (from the left) and their outputs (to the right).

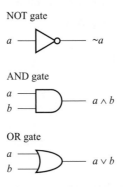

NOT gate

a ———|>o—— $\sim a$

AND gate

a ——
b ——|D—— $a \wedge b$

OR gate

a ——
b ——|D—— $a \vee b$

You may also encounter a NAND gate

a ——
b ——|Do—— $\sim(a \wedge b)$

Figure 4.6

EXAMPLE 4.4

Show whether the combinational circuits shown in figure 4.7 are equivalent by giving the corresponding Boolean expressions and by comparing their truth tables.

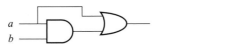

Figure 4.7

SOLUTION

a	∨	(a	∧	b)
0	**0**	0	0	0
0	**0**	0	0	1
1	**1**	1	0	0
1	**1**	1	1	1

a	∧	(a	∨	b)
0	**0**	0	0	0
0	**0**	0	1	1
1	**1**	1	1	0
1	**1**	1	1	1

The columns shown in bold type are the same so the circuits are equivalent.

Note

The absorption rules of Boolean algebra also give these results.

Adders

When adding two binary numbers together the first part of the algorithm is to add the two units bits together. This produces answers of 0 or 1 with or without a carry bit. Subsequent columns involve adding two bits, *plus* the carry bit from the previous column. Combinational circuits can be designed to achieve this, as shown in figure 4.8.

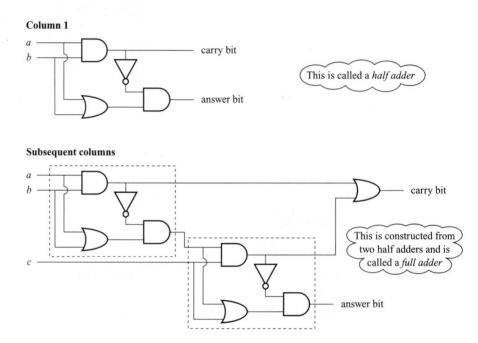

Figure 4.8

The tables of outputs below verify that the combinatorial circuits for the half adder and for the full adder achieve what is required.

Half adder

a	*b*	*carry*	*answer*
0	0	0	0
0	1	0	1
1	0	0	1
1	1	1	0

Full adder

a	b	c	carry	answer
0	0	0	0	0
0	0	1	0	1
0	1	0	0	1
0	1	1	1	0
1	0	0	0	1
1	0	1	1	0
1	1	0	1	0
1	1	1	1	1

EXERCISE 4D

1 Draw switching circuits to represent the following expressions.

(i) $p \vee q \vee r$

(ii) $a \vee (b \wedge c)$

(iii) $(l \vee m) \wedge (l \vee p)$

(iv) $(a \vee b \vee c) \wedge (\sim a \vee b \vee \sim c)$

2 Give Boolean expressions for the following circuits.

(i)

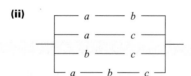

(ii)

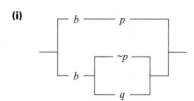

3 Four switches represent the first four binary places.
Wire them so that the light shines when they are set to represent

(i) an even number

(ii) an odd number

(iii) a multiple of three

(iv) a prime number.

4 Each of three members of a committee votes 'yes' or 'no' on a proposal by pressing or not pressing a button. Design a circuit to allow current to pass when and only when at least two of the members vote 'yes'.

5 A light in a room is to be controlled by two switches on separate walls. Flicking any one of the switches is to turn the light on if it is off, and off if it is on. Design a circuit to achieve this.

6 Draw a combinatorial circuit to represent $\sim c \wedge [(a \wedge b) \vee \sim(a \wedge c)]$.

7 Give the Boolean expression corresponding to the following combinatorial circuit.

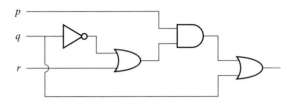

8 A burglar alarm is controlled by a switch. If the switch is off the alarm will not function. If the switch is on then the alarm will function if either the front door, or the back door or both doors are opened. Design a combinatorial circuit to achieve this.

9 Draw a combinatorial circuit which is equivalent to the following switching circuit.

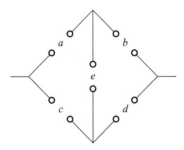

INVESTIGATION

Investigate how to build a combinatorial circuit to subtract one binary number from another.

EXERCISE 4E

1 (i) Show that the proposition $[a \wedge (a \Rightarrow b)] \Rightarrow a$ is a tautology.

(ii) Show that the proposition $(\sim p \vee \sim q) \Rightarrow (p \wedge \sim q)$ always takes the value of p.

(iii) Draw a combinatorial circuit for $a \Rightarrow b$ using OR and NOT gates only.

[AEB]

2 (i) (a) Write down a Boolean expression for the following circuit.

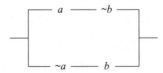

(b) Use the distributive rules to prove that the Boolean expression for the circuit is equivalent to $(a \vee b) \wedge (\sim a \vee \sim b)$.

(c) Use part (b) to draw an alternative, equivalent switching circuit to that given in part (a).

(ii) (a) The following is a combinational circuit for a half adder.

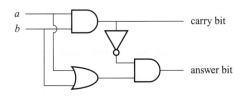

Draw a table of inputs and associated outputs.

(b) The following circuit uses two half adders.

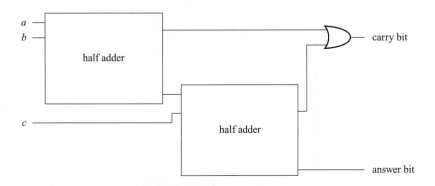

Draw a table of inputs and associated outputs, and explain briefly the purpose of the circuit.

[AEB]

3 (i) Mr C Y Nic claims to be able to prove *anything*. On a fine sunny day he claims to be able to prove that it is raining. His 'proof' is as follows.

Assume that it is raining.

When it rains I carry my umbrella.

I am carrying my umbrella.

At this point Mr Nic waves his umbrella triumphantly, showing the last statement to be true, and shouting, 'There you are – so it is raining!'

Why is this 'proof' not a proof?

(ii) Use a truth table approach to prove that the following two statements are equivalent.

If the north wind blows then we shall have snow.

If we have no snow then the north wind does not blow.

(iii) (a) Write down a Boolean expression for the following circuit.

(b) Show, algebraically or otherwise, that your expression reduces to a ∨ b.

[AEB]

4 (i) A two-way switch is a device which has two connected switches arranged so that when one is open the other is closed, and vice versa. It is convenient to label the two connected parts as (say) *A* and ~*A*. When a two-way switch is operated (flicked) then *A* closed and ~*A* open is changed to *A* open and ~*A* closed, or vice versa.

(a) There are two two-way switches, switch *A* and switch *B*, controlling the light in a room. The circuit below controls the flow of current. Describe what happens when a switch is flicked.

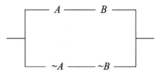

(b) The light in a room is to be controlled in such a way that flicking any one of three separate switches will turn the light on if it is off and off if it is on. Design a circuit which will achieve this.

(ii) Each member of a committee has a button to press in order to register a 'Yes' vote on a proposal. Design *two* different circuits for the buttons for a committee of size three, which will allow current to pass only when two or more members press their buttons.

[AEB]

5 (i) Find a simpler equivalent circuit for the circuit shown below.

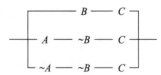

(ii) Argument number 1
If Jones is a Communist, then Jones is an atheist. Jones is an atheist. Hence Jones is a Communist.

Argument number 2
If Smith was not the murderer or if Patel was lying, then Patel did not meet Smith last night. Patel met Smith last night. Hence Smith was the murderer.

For *each* of the two arguments:

(a) replace the elementary propositions by letters (e.g. let *Jones is a Communist* be denoted by *C*), and rewrite the arguments using those letters together with logical connectives ($\sim$, $\wedge$, $\vee$, $\Rightarrow$, $\Leftrightarrow$)

(b) use truth tables to check the validity of both arguments.

[AEB]

6 (i) Use truth tables to prove that the following are true.

(a) $(p \Rightarrow q) \Leftrightarrow \sim(p \wedge \sim q)$

(b) $(\sim q \Rightarrow \sim p) \Leftrightarrow (p \Rightarrow q)$

(ii) Draw a combinatorial circuit to represent the expression
$[a \vee (\sim b \vee c)] \wedge \sim b$.

(iii) Prove that the expression $[a \vee (\sim b \vee c)] \wedge \sim b$ can be simplified to $\sim b$.

[AEB]

7 (i) Use a truth table to prove that $a \vee (b \wedge c)$ is equivalent to $(a \vee b) \wedge (a \vee c)$.

(ii) Express the proposition $p \Rightarrow q$ using p, q and the operations $\sim$ and $\wedge$ only.

(iii) Give the Boolean expression for the output of the following combinatorial circuit.

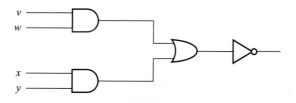

[AEB]

8 (i) Give the Boolean expression corresponding to the combinatorial circuit shown in the diagram.

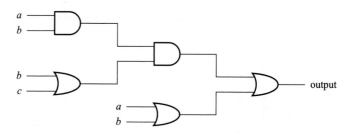

(ii) Use Boolean algebra to simplify your expression from part (i).

(iii) Produce truth tables for your expressions in parts (i) and (ii), and use them to show that the expressions are equivalent.

[AEB]

9 Let *s* represent the proposition *There is snow*.

Let *n* represent the proposition *There is a north wind*.

You are given that if there is no snow then there is no north wind, and that there is a north wind.

(i) Express what you are given in terms of *s*, *n* and logical symbols.

(ii) Use a truth table to prove, from what you are given, that there is snow.

(iii) You are also given that if it snows then the robin hides its head under its wing. What can you deduce about weather and wind if the robin does not have its head hidden under its wing?

[MEI]

10 (i) Draw the switching circuit representing $(a \wedge (b \vee c)) \vee (\sim a \wedge b \wedge c)$.

The diagram shows a circuit for a voting machine for three people, A, B and C. Person A voting for a proposal is represented by *a*. Person A voting against the proposal is represented by $\sim a$.

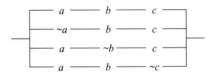

(ii) Show that the expression in part (i) is equivalent to the voting machine in the diagram.

(iii) Draw an equivalent circuit in which the symbols *a*, *b* and *c* are used, twice each, and in which the symbol $\sim$ is not used.

(iv) Draw a circuit for a voting machine in which each of A, B and C has a veto.

[MEI]

11 (i) **(a)** The circuit shown below is called a *half adder*.

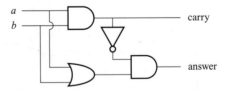

Produce a table showing the outputs, carry and answer, in terms of the inputs, *a* and *b*.

(b) The circuit shown on the opposite page is called a *full adder*.

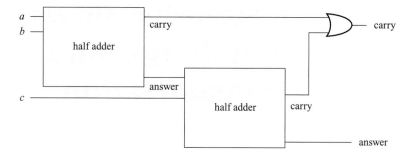

Produce a table showing the outputs in terms of the inputs, a, b and c.

(ii) The cards shown below are from a pack in which every card has a letter on one side and a number on the other.

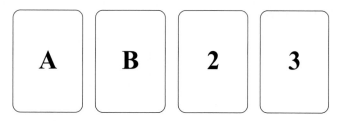

It is claimed that behind a vowel there is always an odd number. Which of the four cards need to be turned over to check this claim?

[MEI]

KEY POINTS

1 Propositions are simple statements which can either be true or false.

2 Compound propositions are constructed using propositions together with the operators $\sim$, $\wedge$, and $\vee$. For convenience $\Leftrightarrow$, $\Rightarrow$ and $\Leftarrow$ are used as well.

3 Truth tables can be used to find the truth values of compound expressions and to prove the validity of statements. A truth table needs 2^n rows, where n is the number of constituent propositions.

4 Propositions, together with the logical operators, form an algebra known as Boolean algebra.

5 Compound propositions can be manipulated using the rules of Boolean algebra. This provides another way of proving statements.

6 Logic is important in information technology because switches in circuits behave in the same way as propositions, and the ways in which switches can be wired together reflect logical operations.

7 Combinatorial circuits are the electronic equivalent of mechanical switching circuits.

5

5 Networks: flows, matchings, allocation and transportation

Oh, could I flow like thee, and make thy stream my great example, as it is my theme!

Cooper's Hill, *Sir John Denham*

Flow networks

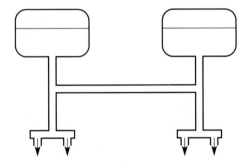

Figure 5.1

Figure 5.1 represents two water cisterns which can deliver water to the four outlets marked by arrows. How much water can be delivered will depend on the sizes of the pipes and the pressure created by the height of the water surface above the point of delivery. In this chapter you will be dealing with some of the mathematics associated with such problems. You will be using capacities and flows in systems of connected pipes, or in similar systems.

There are many physical situations in which something has to be transported from one place to another along alternative routes: water by pipes; electricity by cables; goods by roads; people by trains.

These are modelled by networks in which the weights on the edges are capacities, and in which there are three types of vertex.

The source node, S

There are no flows permitted into this node. All flows start from it. If you are modelling a situation in which there are several sources, for example several warehouses from which goods are dispatched, then introduce a single 'super-source'.

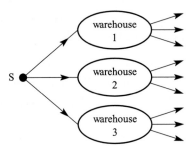

Figure 5.2

The sink node, T

There are no flows permitted out of this node. All flows are into it. If you are modelling a situation in which there are several sinks, for example several shops receiving goods from a distribution network, then introduce a single 'super-sink'.

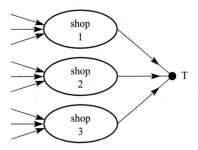

Figure 5.3

All other vertices

Apart from the (super) source and the (super) sink all other vertices have the property that whenever a flow is established, the total flow into the vertex must equal the total flow out.

So figure 5.4 is a network representing the cistern system in figure 5.1.

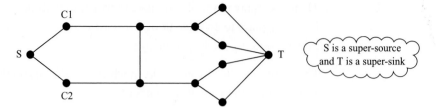

Figure 5.4

Feasible flow

A feasible flow is a set of flows, one for each edge, satisfying the following conditions.

1 No flow may exceed the capacity of an edge.
2 For all vertices other than the source and the sink the total flow into the vertex must equal the total flow out.

Note that in some cases edges will be directed, allowing flows in one direction only, for example edges emanating from the source or edges terminating in the sink. In other cases edges may permit flows in either direction.

❓ Which of the arcs in the network shown in figure 5.4 need to be directed and which can be undirected?

In the network shown in figure 5.5, edge directionalities are marked by arrows where appropriate. Capacities and flows are shown in the boxes associated with each edge.

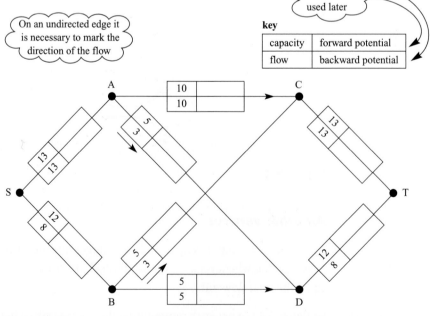

Figure 5.5

❷ Check that the flow pattern indicated is feasible. You will need to check eight edge flows against capacity, for example for SA, $13 \leqslant 13$ so it is feasible. You will also need to check four vertex constraints, for example for vertex A, flow in = 13, flow out = $10 + 3 = 13$ so the constraint is met.

Cuts

The total established flow is 21. You can see this by looking at the flows out of S or into T. You could also see this by imagining cutting through the network so that all the flow from S would 'spill' and no flow would reach T. You could then measure the spill! You can specify such a cut in two ways: by listing the edges which you cut or by splitting the vertices into those on the S side of the cut and those on the T side. The latter method is more systematic and is the approach taken here.

In the network in figure 5.5 there are 16 possible cuts, since for each vertex A, B, C and D there are two possibilities, either it is on the S side of the cut or it is on the T side. For each cut, list all the edges connecting a vertex on the S side with a vertex on the T side. You can then compute both the flow across the cut and the capacity of the cut.

Consider the cut SBD $\mid$ ACT.

There are nine possibilities, SA, SC, ST, BA, BC, BT, DA, DC and DT, but only four of these appear in the network. These are SA, BC, DA and DT.

The flows are 13 in SA, 3 in BC, −3 in DA, and 8 in DT.

Note

A negative flow indicates that the flow is from the sink set to the source set.

These flows sum to 21, which is the total flow through the network. This will be true for *any* cut. The established flow across the cut will be equal to the total flow through the network. Counting the flow across a cut is therefore not very interesting.

Far more interesting is the *capacity* of a cut. This is defined as the sum of the capacities of the edges from the source set to the sink set. Any directed edges leading from sink set to source set are counted as contributing zero. Thus the capacity of the cut SBD $\mid$ ACT is $13 + 5 + 5 + 12 = 35$.

Edge	Capacity
SA	13
BC	5
DA	5
DT	12
Total	35

The capacity of a cut places a limit on the maximum flow which can be sent through a network.

 Calculate the capacity of each of the other 15 cuts.

The maximum flow

The fundamental problem is to find a pattern of flows along the edges of a flow network which maximises the total flow from source to sink. Is the established flow of 21, shown in figure 5.5, a maximum for that flow network?

It looks as though it might be. There is no obvious route from source to sink along which more can be sent. But there *is* in fact one. Such a route is called a *flow-augmenting path*. You need to be able to find such paths and there is a labelling procedure which will help you do this.

The flow-augmenting path is SBCADT. You can send an extra flow of 2 along this path. The presence of CA makes the path look odd but you can effectively put 2 along CA by reducing the existing flow along AC by 2.

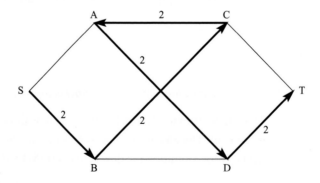

Figure 5.6

The total effect of this is shown in figure 5.7.

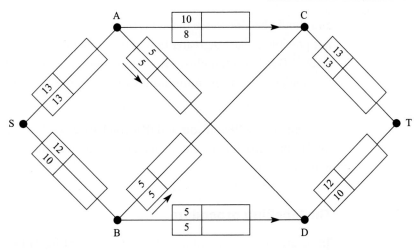

key

capacity	forward potential
flow	backward potential

Figure 5.7

❓ Check that this flow pattern is feasible.

The table below shows the capacities of the 16 possible cuts which you worked out earlier.

Cut	Capacity
S|ABCDT	25
SA|BCDT	27
SB|ACDT	23
SD|ABDT	43
SD|ABCT	42
SAB|CDT	25
SAC|BDT	35
SAD|BCT	34
SBC|ADT	31
SBD|ACT	35
SCD|ABT	60
SABC|DT	23
SABD|CT	27
SACD|BT	42
SBCD|AT	43
SABCD|T	25

❓ Is a flow of 23 the best that can be achieved through this network?

You should be able to see that 23 is the maximum obtainable flow, since it is not possible to have a greater flow across either of the cuts SB | ACDT or SABC | DT. This is an illustration of the *max flow/min cut* theorem, which says that the capacity of the minimum (capacity) cut is the maximum flow which can be achieved.

This implies also that if you find that you have a feasible flow which is equal in total to the capacity of a cut, then that cut is a minimum cut and that flow is a maximum flow.

The labelling procedure

To ensure that you do not miss any flow-augmenting paths you need to label edges to show potential extra flows along edges in *both* directions.

Start from the beginning, with no flows established.

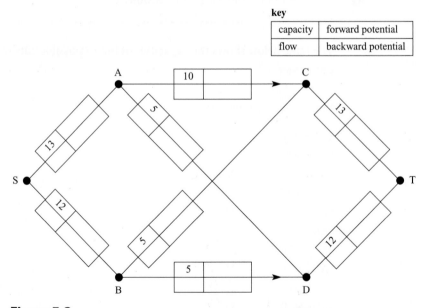

key	
capacity	forward potential
flow	backward potential

Figure 5.8

Since you are starting with no flow, any path will be a flow-augmenting path. Choose a path and find the maximum flow which you can send along it. In figure 5.9 the path chosen is SACBDT, along which you can send 5 units. This flow is labelled, together with potential further flows and potential backflows.

Note that for edges leaving S or arriving at T, backflows are not needed.

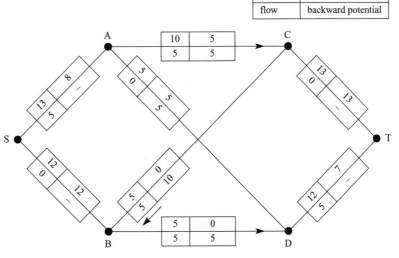

Figure 5.9

For CB the potential backflow is 10 since the existing flow can be reduced to zero and then flow can be pushed in the opposite direction. Thus for undirected edges,

forward potential + backward potential = 2 × edge capacity.

For directed edges, apart from those connected to S or to T

forward potential + backward potential = edge capacity.

Continue looking for paths with spare capacity until there are no more flow-augmenting paths.

Figure 5.10 shows the labelled network after using the flow-augmenting path SACT (capacity 5 units).

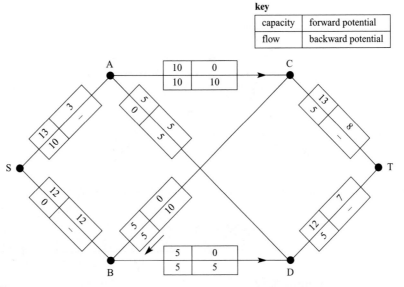

Figure 5.10

Figure 5.11 shows the labelled network after using the flow-augmenting path SBCT (capacity 8 units).

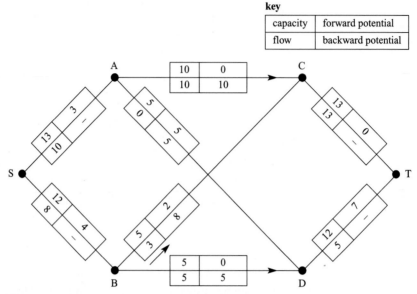

Figure 5.11

Note that before using SBCT there was a flow of 5 from C to B. After using it there is a flow of 3 from B to C. The total change is 8 units of flow, which is a consequence of increasing the flow along SBCT by 8 units.

Figure 5.12 shows the labelled network after using the flow-augmenting path SADT (capacity 3 units).

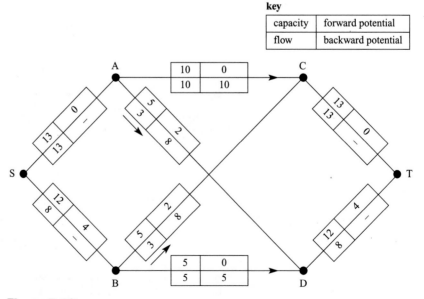

Figure 5.12

Finally, figure 5.13 shows the labelled network after using the flow-augmenting path SBCADT (capacity 2 units). This is easy to find using the labels as it is the only path from S to T with edge potentials greater than zero.

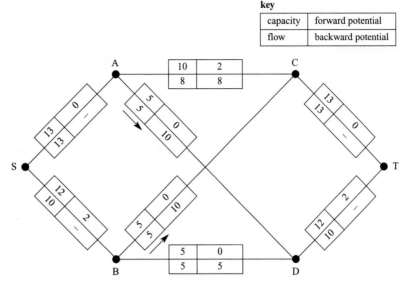

Figure 5.13

You can now see that there is no further flow-augmenting path. For example you cannot move anything more along SA because there is zero potential flow available. You can start along SB but are then blocked by the zero potential on BC and BD.

At this point there will be a cut in which each of the edges has zero potential extra flow.

You can see in figure 5.14 that the edges SA, BC and BD (which form a cut) are at capacity (potential = 0), as are edges CT, AD and BD, which form another cut.

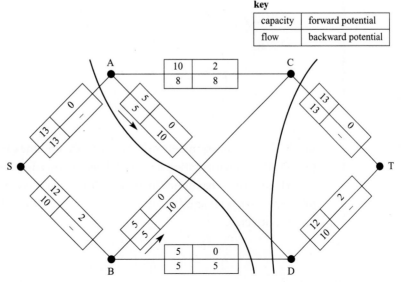

Figure 5.14

Note

In this case we have arrived at the same optimal flow pattern as in figure 5.7. In other cases there may be alternative patterns giving the same optimal total flow.

You will return to network flows in Chapter 6.

1

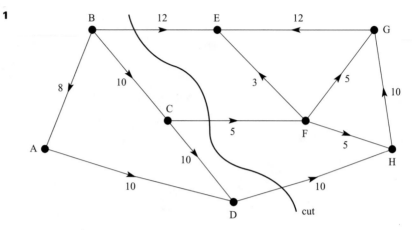

(i) In the transmission network shown above identify the source and the sink.
(ii) Find the capacity of the indicated cut.
(iii) Use flow-augmentation and the labelling procedure to find a flow pattern which gives a maximum total flow.
(iv) Prove that your total flow in part (iii) is maximal.

2

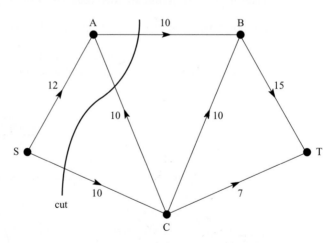

(i) In the transmission network shown above find the capacity of the cut.
(ii) Use flow-augmentation and the labelling procedure to find a flow pattern which gives a maximum total flow. Prove that it is a maximum.
(iii) Repeat parts (i) and (ii) for a similar network in which the edge AC is not directed.

3 Gas is supplied to three locations, T_1, T_2 and T_3, from two sources S_1 and S_2. The rate of supply from S_1 cannot exceed 20 units. The rate of supply from S_2 cannot exceed 30 units. There are no constraints on the rate of flow into T_1, T_2 and T_3. The transmission network is shown in the diagram, pipe capacities giving the maximum permissible rates of flow.

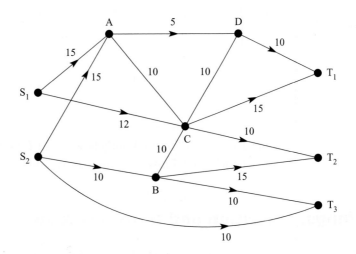

(i) Add a super-source together with appropriate capacitated edges to model the supply constraints.

(ii) Add a super-sink together with appropriate edges.

(iii) Find the maximum total flow through the network, saying how much is delivered to each of T_1, T_2 and T_3.

4 (i) Find minimum cuts for each of these networks.

(a)   (b)

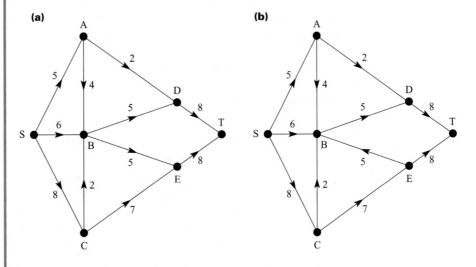

(ii) Confirm that your cuts are minimum cuts by finding a flow of the same value in each network.

5 The matrix represents capacities of roads (thousands of cars per hour) in a road network.

	A	B	C	D	E	F	G
A	–	3	3	–	–	2	–
B	3	–	–	5	1	–	–
C	3	–	–	2	–	3	–
D	–	5	2	–	–	1	2
E	–	1	–	–	–	–	1
F	2	–	3	1	–	–	2
G	–	–	–	2	1	2	–

(i) Draw the network.

(ii) Find the maximum hourly flow of cars from B to F, showing how this may be achieved. Prove that this is a maximum.

Matchings, allocation and transportation

Matching, allocation and transportation problems all share a structure which is modelled by a *bipartite graph*. In this the vertices split into two distinct sets and edges connect a member of one set to a member of the other.

EXAMPLE 5.1

A company is about to reorganise and some staff will have to move. Albert insists that if he is to move, he will only go to the Nottingham office. Beth is more flexible and will go to Leeds, Manchester or Peterborough. Catherine will be happy to be posted to Manchester, Nottingham or Oxford, Desmond to Nottingham or Oxford.

Represent this information on a bipartite graph.

SOLUTION

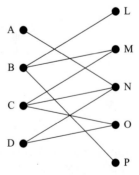

Figure 5.15

Matchings

A *matching* is a subgraph of a bipartite graph satisfying the condition that no vertex is associated with more than one edge.

One possible matching for the staffing problem in Example 5.1 is shown in figure 5.16.

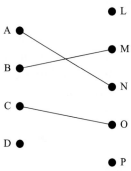

Figure 5.16

 What other matchings can you find?

A *maximal matching* is one with the maximal number of edges. The current matching, shown in figure 5.16, contains three edges. It is clear that you cannot have a matching containing more than four edges, since there are only four members in one of the sets.

You may already have found a matching containing four edges, but the matching improvement algorithm below gives an algorithm for moving from an existing matching to one with a greater number of edges, if one exists.

Step 1   Choose a member of one set which is not in the existing matching. In the staffing problem above this could be D.

Step 2   From D start to construct a tree using edges which are in the graph and which are *not* in the matching. Remember that a tree is a connected graph with no cycles.

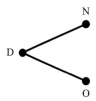

Figure 5.17

Step 3   If a vertex is reached which is not in the existing matching this is 'breakthrough', in which case go to step 6.
In the staffing problem N and O are both in the existing matching, so you do not have breakthrough.

Step 4   Continue the tree using edges which *are* in the matching.

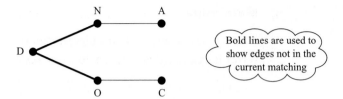

Bold lines are used to
show edges not in the
current matching

Figure 5.18

Step 5   If possible continue the tree using edges which are *not* in the matching
to vertices which are *not* in the tree. If it is not possible conclude that
the initial vertex cannot be brought into the matching.
Go to step 3.

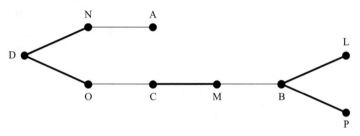

Figure 5.19

Step 6   You now have an path of edges which are alternately not in the current
matching and in the current matching, in this case DOCMBL.

Step 7   Now 'change the state' of each edge in the alternating path. Delete the
edges which were in the matching and bring in those which were not. In
this example edges OC and MB are deleted from the matching and edges
DO, CM and BL are added to the matching as shown in figure 5.20.

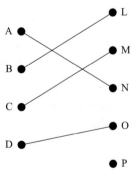

Figure 5.20

Since the first and final edges of an alternating path are not in the current
matching, the result is always an improved matching containing one more edge.

In the staffing example one application of the matching improvement
algorithm produced a maximal matching. In other cases more than one
application may be required.

1 An athletic team has six people to enter for six track events.

      A can run the 100-m and the 200-m.
      B can run the 800-m and the 1500-m.
      C can run the 200-m and the 100-m hurdles.
      D can run the 400-m, 800-m and 1500-m.
      E can run the 400-m and the 800-m.
      F can run the 100-m and the 100-m hurdles.

The captain has pencilled in **A** for the 100-m, **B** for the 800-m, **C** for the 100-m hurdles and **D** for the 400-m. Starting with this matching apply the matching improvement algorithm to produce a complete matching.

2 An over-time schedule is to be produced.

      A can work late on Monday or Wednesday.
      B can work late on Monday or Thursday.
      C can work late on Tuesday, Wednesday or Friday.
      D can work late on Thursday only.
      E can work late on Tuesday or Friday.

One person is required to work late on each day. In how many ways can the manager organise the schedule?

3 A rugby sevens team is to be picked from eight players. The positions in which the players can play are indicated in the table.

		Positions						
		1	2	3	4	5	6	7
	A	✓	✓	✓				
	B			✓	✓	✓	✓	
	C				✓	✓	✓	✓
Players	**D**	✓						
	E			✓	✓	✓		
	F	✓	✓	✓	✓	✓		
	G		✓	✓		✓		
	H			✓	✓	✓	✓	✓

The coach would like to play the following players in the positions given below.

    A–1; B–3; C–4; E–5; G–2; H–6

However, this would leave the team without a full-back (position 7).

(i) Find an alternating path from position 7 to an unused player. Use it to produce an improved matching corresponding to a complete team.

The game is started with the following team.

A–1; B–3; C–7; E–5; F–4; G–2; H–6

Part way through the game, H sustains an injury and D has to come on as substitute.

(ii) Find an alternating path starting from D which reassigns the positions so as to allow D to come on to play in position 1.

4 Consider the bipartite graph shown below.

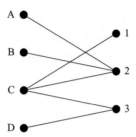

(i) Starting with C connected to 2 and D connected to 3, use the matching improvement algorithm to establish a maximal matching.

(ii) Use flow-augmentation and the labelling procedure to establish a maximal flow through the network shown below. All edges have capacity 1.

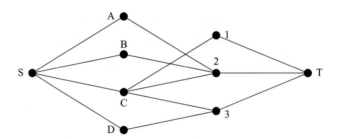

Under what circumstances will your flow pattern imply a solution to the maximal matching problem?

5 The following bipartite graph shows which teachers can teach which subjects.

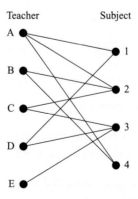

There are 82 possible matchings. Show that there are 8 complete matchings.

Allocation

Members of a 4×400 relay team average the times shown in the table when running in the four positions of the race.

		\ 1	\ 2	\ 3	\ 4
		Race positions			
	A	60	62	61	64
Team	**B**	65	63	60	60
member	**C**	59	61	58	63
	D	60	60	62	60

The problem is to allocate runners to positions so as to minimise the sum of the average times.

The allocation problem is both a specialisation and a generalisation of the matching problem. It is a specialisation in that (usually) the numbers of elements in each set are equal and the bipartite graph is complete, meaning, in this case everybody can do everything. It is a generalisation in that the edges of the graphs have weights.

The objective is to choose the optimal (minimum total weight or maximum total weight) complete matching.

There is a specific algorithm for this problem, the Hungarian algorithm. This algorithm is, however, of limited use since it converts the problem into one which is equally difficult. You will learn how to solve the problem using linear programming, in Chapter 6.

Transportation

In the transportation problem there are locations at which goods are available (warehouses), and locations at which goods are required (shops). The number of warehouses does not have to be the same as the number of shops, and differing quantities of goods may be available at each warehouse and required at each shop.

The total availability of goods must be made equal to the total demand. If it is not then a dummy warehouse is added to 'supply' any shortfall or a dummy shop to mop up any excess.

The table below shows an example in which there are three warehouses and four shops. The costs of transporting an item from each warehouse to each shop are given in the body of the table.

availabilities		9	9	9	7	demands
		S_1	S_2	S_3	S_4	
5	W_1	5	4	7	10	
12	W_2	1	5	4	6	Costs per item transported
17	W_3	11	2	3	5	

143

The problem is to make the deliveries at least cost. The transportation problem is a generalisation of the allocation problem. Again the problem is not solved at this stage. This time there is a specific algorithm which is useful. But again you will solve it using linear programming, since this is how practical problems are almost invariably solved.

INVESTIGATION

Find out about the transportation algorithm. In particular find out about the north-west corner rule and about the stepping stone method for making improvements.

1 The diagram shows a pipe network. The numbers on the arcs give the maximum capacities of the pipes.

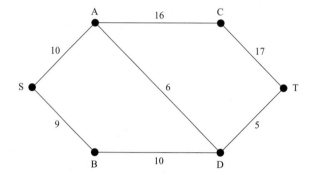

(i) Use the labelling procedure to find the maximum flow from the source S to the sink T. Show and describe the steps of your algorithm.

(ii) Find a cut to prove that your flow is maximal. Explain why it proves that the flow is maximal.

2 The diagram shows the maximum capacities of the edges in a directed distribution network and the flows currently established in the network.

key	
capacity	forward potential
flow	backward potential

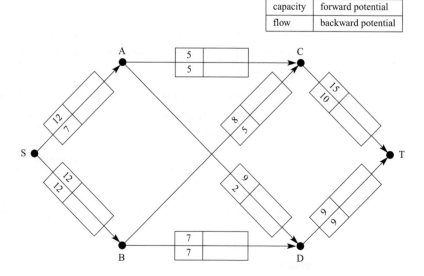

Networks: flows, matchings, allocation and transportation

5

(i) Find a minimum cut and give its capacity.

(ii) Say why the established flows do not give a maximum total flow through the network.

(iii) Find and give a flow-augmenting path, and give a set of flows which do produce a maximum flow through the network.

[AEB]

3 The diagram shows a directed flow network with capacities and established flows.

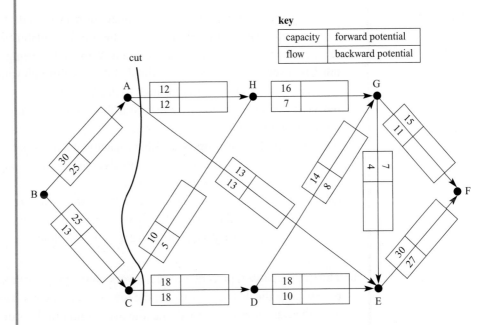

key

capacity	forward potential
flow	backward potential

(i) Identify the source and the sink.

(ii) What is meant by a *cut*? What is the capacity of the given cut?

(iii) Find and indicate one flow-augmenting path, and use it to augment the flow as far as is possible.

(iv) Is the flow now maximal? If not use flow-augmentation to find the maximal flow.
Draw a diagram showing your maximal flow.

(v) Explain how you know that the flow which you quoted in part (iv) is maximal.

[Oxford]

4 The diagram overleaf shows a gas distribution network consisting of three supply points, A, B and C, three intermediate pumping stations, P, Q and R, two delivery points, X and Y, and connecting pipes. The figures by A, B and C are measures of the daily availability of gas at the supply points.

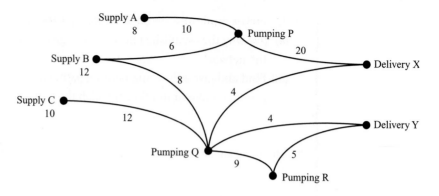

(i) Copy the network, introducing a single source with links to A, B and C, the capacities on the links reflecting the supply availabilities.

(ii) Introduce a single sink linked to X and Y, the links having large capacities.

(iii) Use inspection to find the maximal daily flow through your network, making a list of the flows through each pipe.

(iv) Find a suitable cut to prove that your flow in part (iii) is maximal.

(v) Interpret the flows in the new links.

(vi) A new pipeline is proposed with a capacity of 5 units per day, connecting P and Q. Demonstrate the use of a labelling procedure to augment the flow, and thus find the new maximal flow.

(vii) R is now to become a delivery point. Explain how to adapt the approach of part (ii) to find the maximal daily flow of gas in total that can be delivered to R, X and Y. You do not need to calculate this flow.

[Oxford]

5 The two matrices represent a road network connecting seven towns. The first matrix gives the distances between the towns in kilometres. The second matrix gives the capacities of the roads, i.e. the maximum numbers of vehicles which can pass between towns in one hour (in thousands of cars).

	A	B	C	D	E	F	G
A	–	10	8	–	–	30	–
B	10	–	–	8	15	–	–
C	8	–	–	5	–	7	–
D	–	8	5	–	–	15	4
E	–	15	–	–	–	–	12
F	30	–	7	15	–	–	10
G	–	–	–	4	12	10	–

Distances

	A	B	C	D	E	F	G
A	–	3	3	–	–	2	–
B	3	–	–	6	2	–	–
C	3	–	–	2	–	4	–
D	–	6	2	–	–	2	1
E	–	2	–	–	–	–	2
F	2	–	4	2	–	–	2
G	–	–	–	1	2	3	–

Capacities

(i) Draw the road network.

(ii) Find the maximum hourly flow of cars from B to F, showing how this may be achieved. *Prove* that this is a maximum.

(iii) Use, and demonstrate your use of, Dijkstra's algorithm to find the shortest route between B and F.

(iv) What percentage of the maximum hourly flow of vehicles uses the shortest route?

[Oxford]

6 The network shows a system of pipes. The numbers represent the capacities of the pipes.

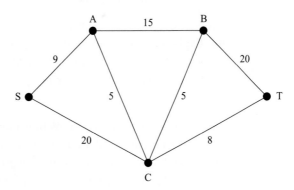

(i) By marking flows in pipes on a copy of the diagram, show that it is possible to achieve a flow of 27 units from S to T.

(ii) Give a cut with a capacity of 27 units. Say what this shows about the flow of 27 units.

(iii) If there must be a flow of at least 2 from A to C in AC what will be the maximum flow from S to T, and what flow will be needed in SC to achieve that maximum?

[AEB]

7 In the following network the edges have capacities in the directions indicated by the arrows. The capacities and established flows are given.

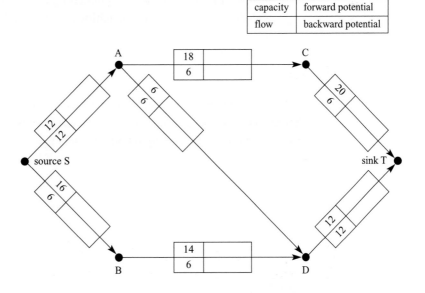

(i) Find a minimum cut and explain why it shows that the existing flow is not a maximum flow. On a copy of the network show the flows along the edges which give the maximum flow from S to T.

Suppose that the flow along the edge AD may now be in either direction.

(ii) Explain why the maximum flow from S to T is now greater than in part (i) and state the value of the new maximum. On a copy of the network show the flows along the edges which give that maximum flow from S to T.

[AEB]

8 As a reward to herself for completing a maths coursework task, Bhavni decides to spend an evening watching programmes on her parents' new digital television. She finds that there are five programmes, A, B, C, D and E, which she would like to watch. Each programme lasts for 1 hour, and they are available at various starting times – 7, 8, 9, 10 and 11 pm. The specific starting times are shown in the table below.

Programme	A	B	C	D	E
Times available	7	8	7	10	9
	8	10	10		11

(i) Construct a bipartite graph to represent this information.

Bhavni's brother says that he had planned to watch programme A at 7 pm, programme E at 9 pm, and programme B at 10 pm, but that he would be happy to watch those programmes at different times if necessary.

(ii) Apply the matching algorithm to the matching specified by Bhavni's brother. Produce an improved matching which still satisfies her brother, and show that a complete matching is not possible.

[MEI, part]

Note

Problems relating to network flows, matchings, allocation and transportation are often solved using linear programming. You will learn more about this in Chapter 6 and will find questions relating to both Chapters 5 and 6 in Exercise 6D.

1 A flow network has a single (super) source and a single (super) sink.

2 A feasible flow has total flow in equal to total flow out for each vertex other than source and sink. It also satisfies the capacity constraints for every arc.

3 A *cut* is a partition of the vertices of a flow network into two sets, one containing the source and the other containing the sink. This defines a corresponding set of arcs, each connecting a vertex in the source set to a vertex in the sink set.

4 The maximum flow possible through a flow network is equal to the capacity of the minimum cut.

5 Arcs in a flow network are labelled with four numbers: capacity, flow, forward potential and backward potential. Labels are updated after the flow is increased using a *flow-augmenting path*.

6 A *matching* is a subgraph of a bipartite graph satisfying the condition that no vertex is associated with more than one edge.

7 A matching can be improved if an *alternating path* can be found which connects vertices which are not in the current matching.

8 *Allocation* and *transportation* problems are variants of the maximal matching problem.

6 Linear programming: applications

Wisdom denotes the pursuing of the best ends by the best means.

Frances Hutcheson

In Chapter 1 you examined the techniques involved in moving from the mathematical formulation of a linear-constrained optimisation problem to the mathematical solution. In this chapter it is assumed that those techniques are implemented in a software package. In this book LINDO is used but most LP packages have similar characteristics.

```
MAX 2X + 3Y
ST  X + Y <= 7
    3X + 5Y <= 30
END

LP OPTIMUM FOUND AT STEP 2

OBJECTIVE FUNCTION VALUE

    1)          18.50000

VARIABLE        VALUE         REDUCED COST
    X         2.500000          0.000000
    Y         4.500000          0.000000

    ROW    SLACK OR SURPLUS    DUAL PRICES
```

Figure 6.1 *Extract from LINDO input and output*

There are two parts to this chapter. In the first part you will look at modelling in linear programming: moving from the real-world problem to the mathematical problem. You will consider a number of example problems.

In the second part you will look at the interpretation of the mathematical solution and at the post-optimal analysis offered by computer packages.

Modelling

Here are some network problems which can be formulated as linear programming problems. They are very simple, but are intended to show *how* to approach such problems.

Shortest path

The following LP finds the shortest path from A to D in the network shown in figure 6.2.

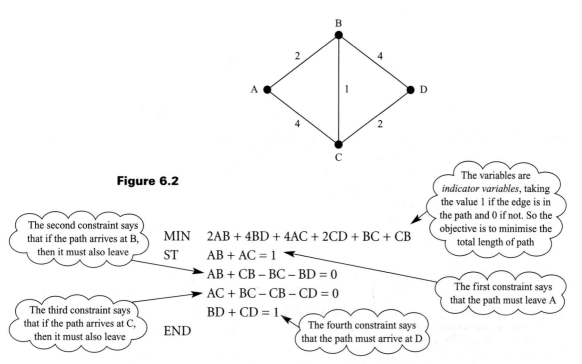

Figure 6.2

The variables are *indicator variables*, taking the value 1 if the edge is in the path and 0 if not. So the objective is to minimise the total length of path

The second constraint says that if the path arrives at B, then it must also leave

MIN   2AB + 4BD + 4AC + 2CD + BC + CB
ST   AB + AC = 1
      AB + CB − BC − BD = 0
      AC + BC − CB − CD = 0
      BD + CD = 1
END

The first constraint says that the path must leave A

The third constraint says that if the path arrives at C, then it must also leave

The fourth constraint says that the path must arrive at D

You might think that you need somehow to insist that the variables be allowed only to take values 0 or 1. Normally this would be the case, and you would then have an *integer programming* problem. Integer programming problems require computational techniques which are much less efficient than the simplex algorithm. However, the structure of this shortest path problem guarantees that the linear programming formulation will automatically give solutions of 0 or 1, so you do not need integer programming. Problems with this property are known as *unimodular* problems.

Note

With a large network, formulating a shortest path problem as an LP would be tedious. It would be quite automatic, but a long process. Since it *is* automatic it would be relatively easy to write computer code to do the task. Again, this is quite common with practical LP problems of many types. Programs which generate the input for an LP are known as *matrix generators*.

 Submit the shortest path LP given above to an LP package, run it and check that your results show that the shortest path is A → B → C → D.

Network flows

Figure 6.3 shows the same network as that in the shortest path problem, except that A has been relabelled S and D has been relabelled T, thus defining a network flow problem.

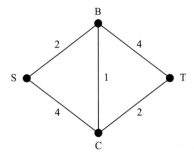

Figure 6.3

An LP which finds the maximum flow from S to T is as follows.

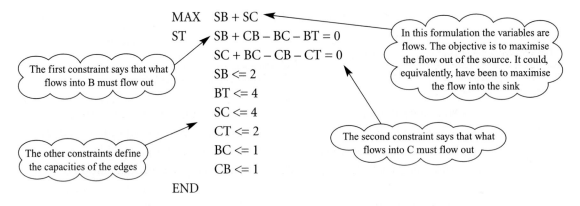

MAX SB + SC
ST SB + CB − BC − BT = 0
SC + BC − CB − CT = 0
SB <= 2
BT <= 4
SC <= 4
CT <= 2
BC <= 1
CB <= 1
END

The first constraint says that what flows into B must flow out

The other constraints define the capacities of the edges

In this formulation the variables are flows. The objective is to maximise the flow out of the source. It could, equivalently, have been to maximise the flow into the sink

The second constraint says that what flows into C must flow out

Note

The simplex algorithm, and most LP packages, use only non-negative variables. Since flow in BC can be in either direction, variables BC and CB have to be used. LINDO allows variables to take negative values if they are declared to be *free*, so if BC is declared to be *free* then CB would not be needed. (Chapter 1 covered how this is achieved by software.)

Submit the network flow LP to an LP package, run it, and check that your results are SB = 2, SC = 3, CB = 1, CT = 2 and BT = 3, giving a total flow of 5 units.

Matching

An LP to achieve a maximal matching from the bipartite graph shown in figure 6.4 is as follows.

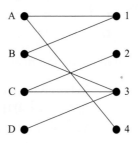

Figure 6.4

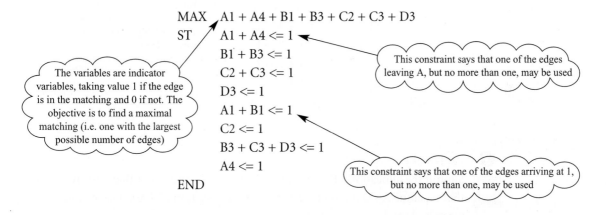

MAX   A1 + A4 + B1 + B3 + C2 + C3 + D3
ST   A1 + A4 <= 1
         B1 + B3 <= 1
         C2 + C3 <= 1
         D3 <= 1
         A1 + B1 <= 1
         C2 <= 1
         B3 + C3 + D3 <= 1
         A4 <= 1
END

The variables are indicator variables, taking value 1 if the edge is in the matching and 0 if not. The objective is to find a maximal matching (i.e. one with the largest possible number of edges)

This constraint says that one of the edges leaving A, but no more than one, may be used

This constraint says that one of the edges arriving at 1, but no more than one, may be used

1 Submit the LP to an LP package, run it, and check that your result is that the following edges should be selected: A4, B1, C2 and D3.

2 Formulate the problem as a network flow problem, using the method shown in Exercise 5B, question 4, and run it.

  Is the matching problem unimodular? (That is, does the structure of the problem guarantee that the solution will be either 0 or 1?)

Allocation

The matrix overleaf shows the costs of performing each of four tasks on each of four machines. Each task has to be performed and each machine must be used.

	Task			
	1	**2**	**3**	**4**
A	5	2	3	6
B	1	7	2	4
C	5	8	3	1
D	4	4	2	6

(**Machine** labels the rows A, B, C, D)

An LP to find the optimum allocation of machine to task is given below.

MIN 5A1 + 2A2 + 3A3 + 6A4 + B1 + 7B2 + 2B3 + 4B4 + 5C1 + 8C2 + 3C3 + C4 + 4D1 + 4D2 + 2D3 + 6D4

ST A1 + A2 + A3 + A4 = 1

   B1 + B2 + B3 + B4 = 1

   C1 + C2 + C3 + C4 = 1

   D1 + D2 + D3 + D4 = 1

   A1 + B1 + C1 + D1 = 1

   A2 + B2 + C2 + D2 = 1

   A3 + B3 + C3 + D3 = 1

   A4 + B4 + C4 + D4 = 1

END

> The variables are indicator variables. For instance, C1 takes the value 1 if machine C is allocated to task 1, and 0 if it is not. It is another unimodular problem. The objective is to minimise the cost of the allocation

---

? Explain the constraints.

 Submit the allocation LP to an LP package, run it, and check that your results are that machine A should do task 2, machine B task 1, machine C task 4, and machine D task 3, at a cost of 6.

Transportation

In this problem the matrix is the same as in the allocation problem, but this time each entry represents the cost of transporting a single item from a warehouse (A, B, C or D) to a shop (1, 2, 3 or 4). To the left of each warehouse is a number indicating the availability of items in stock. Above each shop is a number indicating the requirements of the shop.

		5	5	5	5	**Demands**
	Shop	**1**	**2**	**3**	**4**	
Warehouse						
3	**A**	5	2	3	6	
6	**B**	1	7	2	4	
9	**C**	5	8	3	1	
2	**D**	4	4	2	6	

Availabilities

An LP which solves this transportation problem is given below.

$$\text{MIN} \quad 5A1 + 2A2 + 3A3 + 6A4 + B1 + 7B2 + 2B3 + 4B4 + 5C1 + 8C2 +$$
$$3C3 + C4 + 4D1 + 4D2 + 2D3 + 6D4$$

$$\text{ST} \quad A1 + A2 + A3 + A4 = 3$$
$$B1 + B2 + B3 + B4 = 6$$
$$C1 + C2 + C3 + C4 = 9$$
$$D1 + D2 + D3 + D4 = 2$$
$$A1 + B1 + C1 + D1 = 5$$
$$A2 + B2 + C2 + D2 = 5$$
$$A3 + B3 + C3 + D3 = 5$$
$$A4 + B4 + C4 + D4 = 5$$

$$\text{END}$$

The variables and the objective function are the same as for the allocation problem but the variables are not 0/1 variables

This constraint says that three items must be delivered from warehouse A

This constraint says that five items must be delivered to shop 1

Submit the transportation LP to an LP package, run it, and check that your results are that three items should be sent from A to 2; five items from B to 1, one item from B to 3; four from C to 3, five from C to 4; and two items from D to 2, at a total cost of 38.

Critical path analysis

A critical path in an activity network is a longest route through the network. One way of finding such a route is by changing the shortest route minimisation problem on page 151 into a longest route maximisation problem. You would need to include ≤1 constraints for each variable since no activity can appear more than once in the solution.

This approach is easiest if the activity network has been drawn. The following approach uses linear programming to find the minimum duration for the project and is easy to do without drawing the network. You do, however, need to identify the activities which have no successors, activities C and F in the example below.

The table below shows the activities involved in a construction project.

Activity	A	B	C	D	E	F
Immediate predecessors	–	–	A	A	A, B	D, E
Duration (days)	5	3	10	7	9	2

The following LP finds the minimum duration of the construction project.

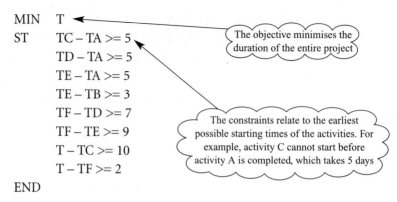

MIN   T
ST   TC – TA >= 5
        TD – TA >= 5
        TE – TA >= 5
        TE – TB >= 3
        TF – TD >= 7
        TF – TE >= 9
        T – TC >= 10
        T – TF >= 2
END

The objective minimises the duration of the entire project

The constraints relate to the earliest possible starting times of the activities. For example, activity C cannot start before activity A is completed, which takes 5 days

Submit the CPA formulation to an LP package, run it, and check that your answer is that the project can be completed in 16 days.

INVESTIGATION

The above formulation will find the duration of a project but cannot identify the critical activities. To do this the latest possible start times must also be considered. Devise a single LP, incorporating both earliest and latest activity start times, to give both the minimum duration of the project and the critical activities.

Maximin problems

There are a number of problems in which the minimum of a set of possibilities is to be made as large as possible, that is, you must maximise the minimum. These are called *maximin problems*. As an example consider a problem from game theory.

Xannos and Yvonne repeatedly play a game in which Xannos secretly writes A or B on a piece of paper, whilst Yvonne secretly writes 1, 2 or 3 on another piece of paper. After they have written their choices they compare what they have written. If the result is A1 then Xannos pays Yvonne 20 pence. The table shows the payments made for this and for each of the other possible results. These are called *returns*. A positive value means that Xannos wins and that Yvonne pays Xannos.

		Y		
		1	2	3
X	A	–20	–10	35
	B	20	0	–15

Xannos cannot make his choice in a predictable way. If his choice is predictable then Yvonne will be able to tailor her response accordingly. For instance, if she knows that he is going to write A then she will write 1. If she knows that he is going to write B then she will write 3. Similarly, Xannos knows that Yvonne's choice will not be predictable. Xannos therefore chooses a tactic in which he writes A randomly with probability p, and B randomly with probability $(1 - p)$.

To find what value of p to choose you first need to find Xannos's expected return when Yvonne chooses each of 1, 2 and 3.

In Chapter 3 it was explained that expected value is calculated by multiplying together each value with the probability of obtaining that value, and then adding those results together. Similarly,

$$\text{Expected return} = \sum_{i}(\text{return}_i \times \text{probability}_i).$$

If Yvonne chooses 1 then Xannos's expected return is

$$-20p + 20(1 - p) = 20 - 40p$$

If Yvonne chooses 2 then his expected return is

$$-10p + 0(1 - p) = -10p$$

If Yvonne chooses 3 then his expected return is

$$35p - 15(1 - p) = 50p - 15$$

The maximin tactic is to choose p so that the smallest of the expected returns is as large as possible. With only one probability to determine this can be done by drawing a graph as shown in figure 6.5.

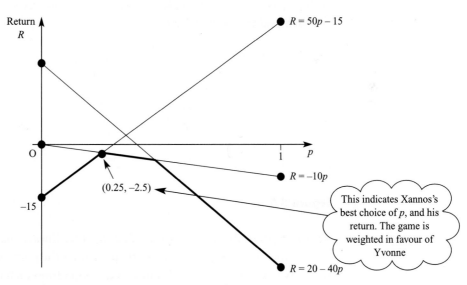

Figure 6.5

For more than two variables an LP is needed. Introduce a new variable M, constrain it to be less than or equal to each return, then maximise it.

Maximise M

Subject to $M \le 20 - 40p$

$M \le -10p$

$M \le 50p - 15$

$p \le 1$

In addition you must allow M to take negative values. The LINDO version of this LP (in which variables must be on the left of the constraint sign) is shown below.

MAX M

ST $M + 40P \le 20$

$M + 10P \le 0$

$M - 50P \le -15$

$P \le 1$

END

FREE M

Submit the maximin LP to an LP package, run it, and check that your solution gives $p = 0.25$ and $M = -2.5$.

Minimax problems

A ship's navigator, in sight of land, takes bearings on three landmarks and marks the back bearings on her chart.

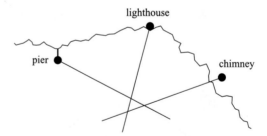

lighthouse

pier

chimney

Figure 6.6

Errors are inevitable in this process, and the three lines do not intersect. She decides that her best guess at the ship's position is the point which has the property that the maximum distance of the point from each of the lines is as small as possible. This is a *minimax problem*.

In a similar problem the lines are defined as follows.

$4x + 3y = 12$

$y = \dfrac{12}{5}x$

$x = 2$

The formula for the distance of the point (p, q) from the line $ax + by + c = 0$

$$= \frac{|ap + bq + c|}{\sqrt{a^2 + b^2}}.$$

So the distance of (p, q) from the line $4x + 3y - 12 = 0$ is either

$$\frac{4p + 3q - 12}{5} \quad \text{or} \quad \frac{-(4p + 3q - 12)}{5}$$

depending on the sign of $4p + 3q - 12$.

You need to introduce a new variable, M, and make M greater than or equal to both $\dfrac{4p + 3q - 12}{5}$ and $\dfrac{-(4p + 3q - 12)}{5}$. It will then certainly be greater than or equal to the positive one.

So an LP formulation for the minimax problem is as follows.

$$\begin{aligned}
\text{Minimise} \quad & M \\
\text{Subject to} \quad & M \geq \frac{4p + 3q - 12}{5} \\
& M \geq \frac{-(4p + 3q - 12)}{5} \\
& M \geq \frac{12p - 5q}{13} \\
& M \geq \frac{-(12p - 5q)}{13} \\
& M \geq p - 2 \\
& M \geq -(p - 2)
\end{aligned}$$

These two inequalities deal with the line $y = \frac{12}{5}x$

These two inequalities deal with the line $x = 2$

You must allow for p and q to be negative and you must do whatever algebraic manipulation is needed for the problem to be accepted by your package.

For LINDO the problem is written as follows.

```
MIN   M
ST    5M – 4p – 3q >= –12
      5M + 4p + 3q >= 12
      13M – 12p + 5q >= 0
      13M + 12p – 5q >= 0
      M – p >= –2
      M + p >= 2
END
FREE  p
FREE  q
```

Submit the minimax problem to an LP package, run it, and check that your solution is that the point is $(1\frac{17}{30}, 2\frac{19}{30})$, with minimax distance of $\frac{13}{30}$. (The package will give the answers as decimal numbers.)

EXERCISE 6A

1 The table shows the distances between the six vertices of a network. Formulate and solve an LP to find the shortest distance from E to B.

	A	B	C	D	E	F
A	0	15	8	5	21	2
B	15	0	11	23	40	17
C	8	11	0	9	26	12
D	5	23	9	0	17	11
E	21	40	26	17	0	13
F	2	17	12	11	13	0

2 The table shows the capacities of links in a distribution network. Formulate and solve an LP to find the maximum flow from S to T and the flows in the links which give that maximum flow.

	S	B	C	D	E	T
S		15	8			
B	15		11	23	4	
C	8	11		9	6	
D		23	9		17	11
E		4	6	17		13
T				11	13	

3 The table shows the costs of allocating individuals A to F to tasks 1 to 6. Given that each individual has to do one task and that each task has to be done by one individual, formulate and solve an LP to find the minimum cost allocation.

	1	2	3	4	5	6
A	10	15	8	5	21	2
B	15	5	11	23	4	17
C	8	11	0	9	6	12
D	5	23	9	30	17	11
E	21	4	6	17	11	13
F	2	17	12	11	13	7

4 The table on the next page shows the cost of sending items from supply points A to F to demand points 1 to 6. The italicised figures to the left of the supply points give the availabilities of items. The italicised figures above the demand points give the requirements for items. Formulate and solve an LP to find the cheapest distribution pattern.

		6 1	32 2	12 3	16 4	10 5	12 6
15	A	10	15	8	5	21	2
23	B	15	5	11	23	4	17
14	C	8	11	0	9	6	12
12	D	5	23	9	30	17	11
24	E	21	4	6	17	11	13
12	F	2	17	12	11	13	7

5 The table gives the returns available to the player of a game. The player can opt to play any one of the alternatives A to F. The opponent can opt to play any one of the alternatives 1 to 6. Formulate and solve an LP to find the player's best strategy.

	1	2	3	4	5	6
A	10	15	8	5	21	2
B	−15	5	11	23	4	17
C	−16	−11	0	9	6	12
D	−8	−3	9	30	17	11
E	−12	−14	6	−17	−11	13
F	2	17	−20	−9	−23	7

6 The table shows, for each of the activities of a project, the immediate predecessors of the task and the duration, in days, of the task. Formulate and solve an LP to find the minimum project duration.

Activity	A	B	C	D	E	F	G	H
Immediate predecessors	–	–	A, B	A	A, B	D, E	A	C, G
Duration (days)	15	10	5	2	8	12	3	13

Indicator variables

There is a very wide range of constrained optimisation problems which require some or all of the variables to take integer values. In some of these situations some or all of the variables are restricted to values 0 or 1. These are called *indicator variables*.

A physical situation in which indicator variables are needed is described below. It shows you how such variables are used.

Fixed or zero costs

Building a factory or owning a car are similar in that they incur a fixed cost. To gain the benefit from producing goods in a factory, that factory will have to be

Linear programming: applications

built or bought. Running a car incurs running expenses and saves fares, but it has to be purchased, and tax and insurance have to be paid to put it on the road.

Suppose that you are to produce x tonnes of a product from Factory X (x could be zero, in which case you will not need the factory). Opening the factory will cost £1 million, and the profit per tonne is £10.

An indicator variable, IX, is used to help to model this. IX can take values 0 or 1. It must be switched on, that is take value 1 if $x > 0$ and switched off, that is take value 0 if $x = 0$.

The objective function will either be a cost function to be minimised, and will include $1\,000\,000IX - 10x$, or will be a profit function to be maximised, and will include $10x - 1\,000\,000IX$.

e   *Switching indicator variables on and off*

 In theoretical terms, switching an indicator variable between values 0 and 1 seems easy to achieve, since you can distinguish between $x = 0$ and $x > 0$. However, there are practical difficulties when using digital arithmetic with limited, albeit large, bit sizes. Digital machines work with approximations and rounding errors creep into calculations. For instance, entering '$=1.000\,000\,000\,000\,001 - 1$' into a cell of Excel 97 leads to the contents of the cell being recorded as 0.

To counter this problem you need to set a threshold for x, beyond which it is deemed to be non-zero.

Suppose that x can take values between zero and some maximum value, M, and let m be a threshold beyond which x is deemed to be non-zero as shown in figure 6.7.

Figure 6.7

Convince yourself by choosing examples for M, m and x, that the following inequalities achieve what is required: $IX \leqslant \frac{x}{m}$ and $IX \geqslant \frac{x - m}{M}$.

In the fixed cost problem you only need to include the constraint $IX \geqslant \frac{x}{M}$, where M is greater than or equal to the largest value which x can take. If x takes any non-zero value this constraint forces, and allows, IX to take value 1. You can allow the optimisation to take care of the threshold. The objective function will force IX to take value 0 if it can, and this will be allowed if $x = 0$. In other situations where you need to use an indicator variable you may need to use a threshold.

Set covering

Another situation which requires indicator variables is set covering.

A private school has advertised for teachers to cover the following subjects: art, home economics, woodwork, music, modern languages, physical education and cultural studies. Teachers are paid according to their experience: part-time teachers are to be paid £250 per week, newly qualified teachers £400 per week and experienced teachers £500 per week.

There are six applicants. Their details are summarised below.

Mrs Ashdown: art and home economics, part-time
Mr Balcombe: woodwork, music, modern languages and cultural studies, newly qualified
Ms Copthorne: home economics, woodwork, modern languages, physical education and cultural studies, experienced
Mr Ditchling: art, music and physical education, part-time
Mr Eastleigh: home economics, modern languages, physical education and cultural studies, newly qualified
Miss Farnham: art, woodwork and physical education, part-time

A formulation to find the least-cost solution to cover all subjects is given below.

$$
\begin{aligned}
&\text{MIN} \quad 250A + 400B + 500C + 250D + 400E + 250F \\
&\text{ST} \quad\;\; A + D + F \geq 1 \\
&\qquad\quad A + C + E \geq 1 \\
&\qquad\quad B + C + F \geq 1 \\
&\qquad\quad B + D \geq 1 \\
&\qquad\quad B + C + E \geq 1 \\
&\qquad\quad C + D + E + F \geq 1 \\
&\qquad\quad B + C + E \geq 1 \\
&\text{END} \\
&\text{INT} \quad\; 6
\end{aligned}
$$

Mrs Ashdown, Mr Ditchling and Miss Farnham can teach art. This constraint says that art must be taught

This last statement is LINDO syntax for declaring 0/1 variables. It says that the first six variables encountered, which in this case is all of the variables, will be 0/1 variables

Submit the problem to an LP package which can deal with 0/1 variables, run it, and check that your solution is to employ Ms Copthorne and Mr Ditchling at a cost of £750 per week.

You might imagine that it is relatively easy to adapt an LP package to cope with 0/1 variables. If, for instance, variable x is defined to be an indicator variable then solve the LP with $x = 0$, solve it again with $x = 1$, then choose the better solution. But if there are n indicator variables this approach would require the solutions to 2^n LPs, and this can quickly become unmanageable as n increases. There are more efficient approaches, but they are not covered in this book.

Integer programming

In this section you will see how to generalise the set covering problem to the set packing (knapsack) problem, a simple example of a full integer programming problem.

Items A, B, C, D and E (as many as required of each) are available to be packed into a crate. The weight of the packed crate must not exceed 30 kg. The weight and value of each item is shown in the table.

Item	A	B	C	D	E
Weight (kg)	2	7	5	8	3
Value (£)	2	10	6	11	4

An integer programming formulation to maximise the value of packed items is:

MAX   2XA + 10XB + 6XC + 11XD + 4XE

ST   2XA + 7XB + 5XC + 8XD + 3XE <= 30

END

GIN   5 ◄— This last line is a LINDO statement which declares that there are five integer variables. GIN stands for general integer (variable)

Submit this problem to a mathematical programming package which has integer programming capability, run it, and check that your solution is to pack two of item B and two of item D, with a value of £42.

You need to be careful with $<$ or $>$ constraints when you are using integer programming. Since you are dealing with linear programming the mathematical assumption is that the variables can take real number values. In practice, LP problems are solved on computer systems, which operate with rational numbers. Suppose a crude computer operates with four-digit accuracy and suppose you wished to model the constraint $x < 4$. Would you allow the machine to accept values of x up to 3.999, or values up to $x = 4$? If you choose the former then all values greater than 3.999 and less than 4 would be mistakenly rejected; if the latter then only the single value $x = 4$ would be mistakenly accepted. Thus LP computer packages generally draw no distinction between $x \leqslant 4$ and $x < 4$. Thus if x is an integer and $x < 4$, a package may well allow $x = 4$.

EXERCISE 6B

1 Formulate LPs to find the maximum flows in the networks in Exercise 5C, questions 1 and 2.

2 Formulate LPs to find the maximum flow from B to F in Exercise 5C, question 5 part (ii) and the shortest distance from B to F in part (iii) of the same question.

3 Formulate LPs to solve the maximal matching problems in Exercise 5B questions 1 and 4 part (i).

4 Use linear programming to solve the allocation problem on page 143 and the transportation problem on pages 143–144.

5 A contractor needs to move 570 tonnes of landfill over a three-day period. He can hire 10-tonne lorries and/or 15-tonne lorries and/or 20-tonne lorries on each day. A 10-tonne lorry costs £100 per day plus 50p per mile. A 15-tonne lorry costs £120 per day plus 65p per mile. A 20-tonne lorry costs £150 per day plus 90p per mile. A round trip from the quarry to the landfill site is 20 miles and each lorry can make five trips per day. Find the minimum-cost solution.

(Hint: The three-day period is irrelevant, at least as far as the question is concerned. You only need to know for how many days each type of lorry is hired. (One 10-tonne lorry for two days is the same as two 10-tonne lorries for one day.) Use integer variables for the number of days each type of lorry is hired. You will also need integer variables for the number of trips made by each type of lorry.)

Post-optimal analysis

The solution to a linear programming problem provided by a computing package includes more than just the answer. Much information can be extracted from the final values recorded by the algorithm. Not all of this information is relevant to all applications. For instance, one part of the post-optimal analysis gives the benefits of relaxing each of the constraints, but it is not relevant to know the value of relaxing a constraint when modelling a network flow problem. The constraints model the fact that flows into a vertex must be balanced by flows out, and it is not possible to relax such constraints.

The information provided is known as the post-optimal analysis. You can study it by applying LINDO to the drinks production problem in Chapter 1, where most of the post-optimal analysis *is* relevant.

The problem was to produce x litres of energy drink and y litres of refresher drink to maximise income subject to constraints on the availability of syrup, vitamin supplement and flavouring. The problem was formulated as follows.

MAX	$x + 0.8y$	⟵ income
ST	$x + y <= 1000$	⟵ syrup constraint
	$2x + y <= 1500$	⟵ vitamin constraint
	$3x + 2y <= 2400$	⟵ flavouring constraint
END		

The output from LINDO is shown in figure 6.8.

```
LP OPTIMUM FOUND AT STEP 2
              OBJECTIVE FUNCTION VALUE
        1)     880.0000

        VARIABLE          VALUE         REDUCED COST
          X            400.000000         0.000000
          Y            600.000000         0.000000

        ROW      SLACK OR SURPLUS    DUAL PRICES
        2)          0.00000000         0.400000
        3)        100.000000           0.000000
        4)          0.00000000         0.200000

NO. ITERATIONS = 2
RANGES IN WHICH THE BASIS IS UNCHANGED:
                  OBJ COEFFICIENT RANGES
        VARIABLE       CURRENT          ALLOWABLE        ALLOWABLE
                         COEF           INCREASE         DECREASE
          X            1.000000         0.200000         0.200000
          Y            0.800000         0.200000         0.133333

                  RIGHTHAND SIDE RANGES
        ROW            CURRENT          ALLOWABLE        ALLOWABLE
                         RHS            INCREASE         DECREASE
        2)           1000.000000       200.000000       100.000000
        3)           1500.000000       INFINITY         100.000000
        4)           2400.000000       100.000000       400.000000
```

Figure 6.8

The solution is clear; make 400 litres of the energy drink and 600 litres of the refresher drink, giving an income of £880, as you can see from the first blocks of the output.

In this case both variables have non-zero values. In a case in which a variable has zero value the reduced cost gives the amount by which its coefficient in the objective function would have to change to make it worth bringing it into the solution. In a cost minimisation problem the objective function is a cost function and the coefficients of the variables are costs. If the cost of a variable is high it will not appear in the solution, hence the terminology. In a profit maximisation problem 'reduced cost' is interpreted as 'increased profit'. It gives the amount by which the coefficient will have to increase to bring the variable into the solution.

The next block of information relates to the constraints. There is a row for each constraint. The value in the first column gives the extent to which there is any slack left in $\leqslant$ constraints or surplus in $\geqslant$ constraints. Thus the solution shows that there is some vitamin supplement left. In the original problem the vitamin constraint, $\frac{2}{5}x + \frac{1}{3}y \leqslant 300$, was simplified to $2x + y \leqslant 1500$ to make the graphical approach easier.

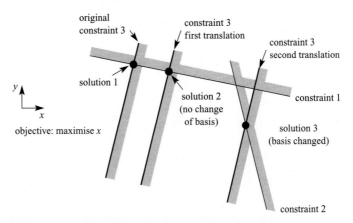

Had $0.4x + 0.2y \leqslant 300$ been used in the computer formulation the slack would have been 20, and that is how many units of vitamin supplement are left over.

'Dual price' gives an indication of how much it is worth to relax a constraint. Again the earlier manipulation of the inequalities makes it more difficult to interpret these. The original syrup constraint was $0.25x + 0.25y \geqslant 250$ which was simplified to $x + y \leqslant 1000$. Had the original constraint been used, the dual price would have been 1.6 and this tells you that, for small changes, every extra litre of syrup will bring in an extra £1.60 in income. It thus gives the marginal value of syrup, which you can compare with the cost.

Replace the first constraint, $x + y \leqslant 1000$, with $0.25x + 0.25y \leqslant 251$, and run the LP. Check that the income is increased to £881.60.

This analysis of dual prices is valid only for changes that do not fundamentally alter the nature of the solution. A change to the right-hand side of a constraint is equivalent to translating the corresponding boundary element (line, plane or hyperplane). The marginal value analysis is valid as long as such a translation leaves the translated optimal point optimal. If not, then it is said that a change of basis has occurred.

In figure 6.9, solutions 1 and 2 are given by the intersection of constraint 3 and constraint 1. Solution 3 is given by the intersection of constraint 3 and constraint 2. This is what is meant by a change of basis.

This leads to the final part of the post-optimal analysis. Remember that the objective function is $x + 0.8y$. The first part of this output shows that the coefficient of x is 1 but that it can go up as far as 1.2 or down as far as 0.8 without the basis changing. You might find this useful, for instance, if looking for alternative solutions to shortest path, flow or matching problems since zero allowable change would indicate the existence of an alternative solution. In geometric terms, altering the cost coefficient corresponds to altering the gradient of the objective function.

The right-hand side ranges give the limits of the constraint translations which can be made without changing basis.

Figure 6.9

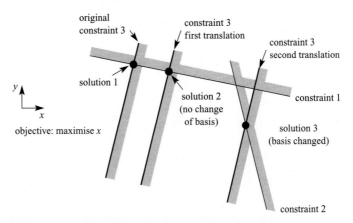

EXERCISE 6C

1 Use an LP package to solve Exercise 1D, question 6 and perform a post-optimal analysis on the solution.

2 Repeat for Exercise 1D, question 8.

3 Repeat for Exercise 1D, question 12.

INVESTIGATION

Explore the use of post-optimal analysis in shortest path, maximum flow, matching, allocation and transportation problems.

EXERCISE 6D

1 The following table gives the mean times (in seconds) over a number of trials taken by four runners to complete each of the four legs in a 4×400 metres relay race.

		Leg			
		1	2	3	4
	A	66	62	66	61
	B	66	65	64	60
Runner	C	63	66	64	62
	D	66	66	65	65

Use linear programming to find the best ordering for these four runners in a competitive race.

Give the best order and the expected total time.

2 The linear program below is to solve a shortest path problem in the graph shown. The structure of the problem ensures that, in the solution, each x value is either 1 or 0, indicating whether or not the corresponding edge is in the solution.

For example, $x_{AB} = 1$ shows that the edge AB is in the solution.

Minimise $P = 2x_{AB} + 4x_{AC} + x_{BC} + x_{CB} + 8x_{BD} + 5x_{CD}$
subject to $x_{AB} + x_{AC} = 1$
$x_{AB} + x_{CB} - x_{BC} - x_{BD} = 0$
$x_{AC} + x_{BC} - x_{CB} - x_{CD} = 0$
$x_{BD} + x_{CD} = 1$
all x-values $\geqslant 0$

The solution given by a computer package is:

P	x_{AB}	x_{AC}	x_{BC}	x_{CB}	x_{BD}	x_{CD}
8	1	0	1	0	0	1

(i) Give the path indicated by the solution, and its length.
The four equality constraints refer to vertices A, B, C and D respectively.
(ii) Explain briefly the purpose of each constraint.

[AEB]

3 The annual cost of supplying six warehouses from six depots is £8000 plus costs specified in the following matrix (each warehouse is to be supplied from exactly one depot; each depot is to supply exactly one warehouse).

£ hundreds per annum		Warehouses					
		1	2	3	4	5	6
Depots	A	6	5	3	0	1	2
	B	1	2	5	0	3	0
	C	10	7	0	2	0	3
	D	0	0	0	4	1	5
	E	6	3	2	0	5	4
	F	3	7	0	3	2	1

Apply linear programming to find the minimal cost solution.

4 A contractor hires out heavy equipment such as cranes. There are eight jobs that he has contracted for. Each job should take less than half a day but the sites are separated and take some time to reach.

The table gives the journey times to or from the main depot and the job durations.

Job	A	B	C	D	W	X	Y	Z
Journey time (hours)	2	1	1	3	2	2	4	1
Duration of job (hours)	4	4	4	3	4	2	4	5

Any of jobs A, B, C or D can be followed by any of jobs W, X, Y or Z, but the same machine cannot do more than one out of each of these groups, so that four machines will need to be used. The travel times between jobs that can be shared are given in the table.

Job	W	X	Y	Z
A	1	2	2	3
B	2	1	3	2
C	1	4	3	3
D	3	2	1	1

(i)   If one machine were to be assigned to do jobs A and W, for how long would it be away from the depot?

(ii) Use linear programming to decide how to pair jobs to be done by one machine, making clear the criterion that you are using.

(iii) What is the longest time that any machine is away from the depot?

5 The table overleaf shows, for each of the activities of a project, the immediate predecessors of the task and the normal duration, in days, of the task. Extra resources can be employed to speed up activities as indicated in the final two rows of the table.

Activity	A	B	C	D	E	F	G	H
Immediate predecessors	–	–	A, B	A	A, B	D, E	A	C, G
Normal duration (days)	15	10	5	2	8	12	3	13
Minimum duration (days)	13	7	3	2	7	9	2	12
Cost of acceleration (£ per day)	1000	750	1500	–	1000	250	500	1500

Formulate and solve an LP to find the minimum project duration and the least additional cost involved in achieving that minimum. List which activities need to be speeded up, and by how much.

Hint: Complete the formulation which has been started for you. The 10 000 is arbitrary. It is there to place a high cost on the time taken so that the project time is shortened as much as possible.

MIN $10000T + 1000SA + 750SB\ldots$

ST $TC - TA + SA >= 15$

      $\ldots$

      $SA <= 2$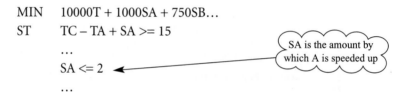

SA is the amount by which A is speeded up

      $\ldots$

6 An electronics company has television manufacturing plants in cities X and Y. The company is planning to produce some components in cities A, B and C, where wages are lower.

The total requirements for the components during the next five years are shown below.

Television plant	Components required
X	30 million
Y	10 million

The initial costs of building production plants as well as the component production costs and plant capacities, over five years, are shown below.

Component plant	Initial building cost (£)	Component production cost (£)	Production capacity
A	20 million	3.00	15 million
B	40 million	2.00	25 million
C	30 million	2.50	20 million

The unit costs, in pounds, of shipping components to X and Y are shown below.

		To X	Y
	A	0.50	1.00
From	B	0.40	1.20
	C	0.20	0.80

The company wants to meet its component needs for the next five years at the minimum total cost, and must decide where it should produce the components.

This problem is formulated as a mixed linear and integer programming problem, with AX representing the number of components supplied from A to X, and similarly for BX, CX, AY, BY, and CY. Indicator variables, ZA, ZB and ZC are used. ZA is forced to take value 1 if a production plant is established in city A, and 0 otherwise, and similarly for ZB and ZC.

(i) The objective function includes the three terms $20\,000\,000$ZA, 3.5AX and 4AY. Explain each of these terms and write out the complete objective function.

(ii) The constraints include one of the form $AX + AY - k$ZA $\leqslant 0$. Explain the purpose of this constraint.
What values are possible for k?
Write down two similar constraints relating to B and to C.

(iii) Write down the constraints which ensure that the requirements at X and Y are met.

(iv) Solve the problem using your computer package. Interpret the solution to the problem given by your output.

7 (i) (x_1, y_1) and (x_2, y_2) are two fixed distinct points in a plane. λ_1 and λ_2 are variables.
Describe the set of points (x, y) given by:

$$x = \lambda_1 x_1 + \lambda_2 x_2$$
$$y = \lambda_1 y_1 + \lambda_2 y_2$$
$$\lambda_1 + \lambda_2 = 1$$
$$0 \leqslant \lambda_1 \leqslant 1$$
$$0 \leqslant \lambda_2 \leqslant 1.$$

(ii) (x_1, y_1), (x_2, y_2) and (x_3, y_3) are three fixed distinct points in a plane. λ_1, λ_2 and λ_3 are variables.
Describe the set of points (x, y) given by:

$$x = \lambda_1 x_1 + \lambda_2 x_2 + \lambda_3 x_3$$
$$y = \lambda_1 y_1 + \lambda_2 y_2 + \lambda_3 y_3$$
$$\lambda_1 + \lambda_2 + \lambda_3 = 1$$
$$0 \leqslant \lambda_1 \leqslant 1$$
$$0 \leqslant \lambda_2 \leqslant 1$$
$$0 \leqslant \lambda_3 \leqslant 1.$$

(iii) (x_i, y_i), $i = 1, 2, ..., n$ are n fixed distinct points in a plane. $\lambda_1, \lambda_2, ..., \lambda_n$ are variables.

Describe the set of points (x, y) specified by the following linear program.

For a given x, the corresponding value of y is:

$$y = \text{minimum} \quad \left(\sum_{i=1}^{n} \lambda_i y_i \right)$$

$$\text{subject to} \quad \sum_{i=1}^{n} \lambda_i y_i = x$$

$$\sum_{i=1}^{n} \lambda_i = 1$$

$$0 \leqslant \lambda_i \leqslant 1 \text{ for } i = 1, 2, ..., n.$$

[Oxford]

8 The formula for the distance from the point (p, q) to the line $ax + by + c = 0$ is

$$\frac{|ap + bq + c|}{\sqrt{a^2 + b^2}}.$$

(i) Find the distance from the point $(3, 4)$ to the line $3x + 4y = 20$.
Find the distance from the point $(1, 1)$ to the line $3x + 4y = 20$.
Explain the significance of the modulus sign in the formula for the distance of a point from a line.

Consider the following linear programming problem.

$$\begin{aligned} \text{Minimise} \quad & z \\ \text{subject to} \quad & z \geqslant 4 - 0.6x - 0.8y \\ & z \geqslant x \\ & z \geqslant y \end{aligned}$$

(ii) This LP is to find the point whose maximum distance from each of three lines is minimised (a minimax problem).
What are the equations of the three lines?

(iii) An LP package gave the following answer to the problem.

$$x = 1.666\,666\,667$$
$$y = 1.666\,666\,667$$
$$z = 1.666\,666\,667$$

Interpret the LP solution in terms of the minimax problem and produce a diagram illustrating your interpretation.

(iv) Explain why the solution point is equidistant from all three lines. Will this always be the case, whatever three lines are used?

(v) In a more general application of the approach, for instance with more than three lines, it would be necessary to include pairs of inequalities such as

$$z \geqslant 4 - 0.6x - 0.8y$$
$$-z \leqslant 4 - 0.6x - 0.8y.$$

Explain why this is so, using diagrams to illustrate your explanation.

[Oxford]

9 There are six cities in Kilroy County and the authorities need to determine where to build fire stations. They want to build the minimum number needed to ensure that at least one fire station is within 15 minutes driving time of each city. The times in minutes needed to drive between the cities are shown below.

	City 1	City 2	City 3	City 4	City 5	City 6
City 1	0	10	20	30	30	20
City 2	10	0	25	35	20	10
City 3	20	25	0	15	30	20
City 4	30	35	15	0	15	25
City 5	30	20	30	15	0	14
City 6	20	10	20	25	14	0

Given that fire stations are to be built in cities, formulate an integer programming problem that will help to decide where to build them.

10 A construction company has a contract for disposing of $200\,000\,\text{m}^3$ of material during the construction of a by-pass. The material has to be moved over a period of 100 working days.

The company intends to hire a fleet of tipper lorries for this contract. Three sizes of lorry are available: large lorries with a capacity of $4\,\text{m}^3$, medium lorries with a capacity of $2.4\,\text{m}^3$, and small lorries with a capacity of $2\,\text{m}^3$. Each lorry can make 10 trips per working day.

The daily costs of hiring are £80 for a large lorry, £40 for a medium lorry, and £32 for a small lorry.

The company has 65 drivers available. To maintain flexibility in disposing of the material the company wishes to have at least 20 small lorries available.

(i) Formulate an integer linear programming problem to decide how many of each type of lorry the company should hire so as to minimise the total hiring cost.

(ii) Use your linear programming software to solve the problem, and interpret your solution.

(iii) Find the cost of the requirement that 20 small lorries should be available.

As an alternative to hiring, the company can use some of its own lorries. The daily cost would be 75% of the cost of hiring for each type of lorry, but it would be necessary to establish a local depot for the duration of the contract. This would cost £50 000, and would allow the company to use up to 50 of its own lorries. Again, at least 20 small lorries would need to be available.

(iv) Adapt your integer linear programming formulation to model this situation. (Use six general integer variables and one 0/1 indicator variable. The indicator variable should be set to 1 if any of the company's own lorries are used.)

(v) Solve the problem and interpret your solution. Determine whether or not there are any cost savings to be made by opening a depot.

[MEI]

11 An island has three electricity power stations. Their maximum power outputs are 4, 5 and 7 MW (megawatts) respectively. Each can be operated at any power output up to its maximum. Their respective hourly costs are 0.75, 0.70 and 0.73 monetary units per MW.

(i) Explain why the following linear program will find the best way of fulfilling an hourly demand for 6.5 MW.

Minimise $\quad 3x_1 + 3.5x_2 + 5.11x_3$
subject to $\quad$ demand $= 6.5$
$\qquad\qquad 4x_1 + 5x_2 + 7x_3 - \text{demand} = 0$
$\qquad\qquad 0 \leqslant x_1 \leqslant 1$
$\qquad\qquad 0 \leqslant x_2 \leqslant 1$
$\qquad\qquad 0 \leqslant x_3 \leqslant 1$

(ii) Use your linear programming package to solve the problem. Include a printout of the solution and interpret that solution.

(iii) Find the best solution to satisfy an hourly demand for 4.7 MW.

(iv) A larger island has 10 power stations, with maximum power outputs and hourly costs per MW as follows.

Station number	1	2	3	4	5	6	7	8	9	10
Maximum power (MW)	4	5	7	6	4	3	8	4.5	6.2	9
Hourly costs per MW	0.75	0.70	0.73	0.72	0.76	0.69	0.77	0.74	0.76	0.73

Use your linear programming package to find the cheapest way to satisfy an hourly demand for 38.2 MW.

(v) A new power station is to be constructed on the larger island. This will cost 4.7 units per hour to run, plus an hourly cost of 0.24 units per MW. Its capacity will be 10 MW. Incorporate this power station into your electricity supply model for the larger island and find the best solution for an hourly demand of 38.2 MW.

(vi) By using your LP model or otherwise, find the minimum hourly demand for power for which it is worth using the new station. You are given that this minimum number of MW is an integer.

[MEI]

12 A type of plastic sheeting is produced in rolls which are 7 m wide. These are cut into rolls of smaller width, as shown in the diagram.

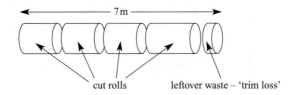

cut rolls   leftover waste – 'trim loss'

(i) On a particular day there is a need for 150 rolls of width 4 m, 200 rolls of width 3 m and 100 rolls of width 2 m. To provide these, full-sized rolls can be cut according to the five plans shown in the table.

Cutting plan	4 m rolls	3 m rolls	2 m rolls	Trim loss (m)
1	1	1	0	0
2	1	0	1	1
3	0	2	0	1
4	0	1	2	0
5	0	0	3	1

(a) The following LP is formulated to satisfy the requirements whilst minimising production.

Minimise $x_1 + x_2 + x_3 + x_4 + x_5$
subject to $x_1 + x_2 \geqslant 150$
               $x_1 + 2x_3 + x_4 \geqslant 200$
               $x_2 + 2x_4 + 3x_5 \geqslant 100$

end
gin   5

Explain what the variables represent and the meaning of the constraints. Run the LP. Produce a printout of your results and interpret those results.

(b) Run the following LP and explain what it is achieving.

Minimise $x_2 + x_3 + x_5$
subject to $x_1 + x_2 \geqslant 150$
               $x_1 + 2x_3 + x_4 \geqslant 200$
               $x_2 + 2x_4 + 3x_5 \geqslant 100$

end
gin   5

(ii) On another day the demand is for 150 rolls of width 4 m, 100 rolls of width 3 m, 250 rolls of width 2 m and 300 rolls of width 1.6 m.

(a) Copy and complete the table below, showing efficient cutting plans (that is, plans with a trim loss < 1.6 m).

Cutting plan	4 m rolls	3 m rolls	2 m rolls	1.6 m rolls	Trim loss (m)
1	1	1	0	0	0
2					
3					
4					
5					
6					
7					
8					
9					
10					
11	0	0	0	4	0.6

(b) Formulate, run and interpret an LP to find how many full rolls should be cut using each cutting plan. How much trim loss is created?

[MEI]

13 Virginia is setting up an airline. She has a capital budget of $500 million ($500m) to buy aeroplanes.

Large planes cost $18m each and have a capacity of 250 passengers. They incur fixed costs of $3m each per annum, plus variable costs of $2.9 per mile. Medium-sized planes cost $15m each and have a capacity of 200 passengers. They have fixed costs of $3m per annum and variable costs of $2 per mile. Small planes cost $12m each and have a capacity of 150 passengers. Their fixed costs are $1.5m per annum and their variable costs are $2 per mile.

Virginia's company will fly long-haul routes, transatlantic routes and short-haul routes. The average distances and demands for these routes are as shown in the table below.

	Distance (miles)	Demand (million passengers per year)
Long-haul	8000	0.5
Transatlantic	5000	1
Short-haul	1000	2.25

Each plane will fly for 300 days a year. Each plane can make one long-haul flight per day, or two transatlantic flights per day, or four short-haul flights per day.

Let NL be the number of large planes, LL the number of long-haul flights using large planes, TL the number of transatlantic flights using large planes and SL the number of short-haul flights using large planes. Use similar variables for medium and small planes.

(i) Explain why the annual cost of using large planes is given by

$$3\,000\,000NL + 23\,200LL + 14\,500TL + 2900SL.$$

(ii) Explain why the inequality

$$1LL + 0.5TL + 0.25SL \leqslant 300NL$$

models the availability of large planes.

(iii) Formulate an LP to find how many planes of each type Virginia should buy so that she can satisfy demand within her capital budget at minimum annual cost.
- For each type of plane you will need an availability inequality.
- For each distance category you will need an inequality to ensure that there is sufficient capacity.
- You will need an inequality to ensure that the capital budget is not exceeded.

(iv) Run your LP, and interpret your solution.
(v) Criticise the model.

[MEI, part]

14 Bhavni plans a 12-hour TV binge starting at 1 pm one day. The programmes which she plans to watch each last for 1 hour. The starting times at which the programmes are available are shown in the table below.

Programme	M	N	O	P	Q	R	S	T	U	V	W	X
Times available	2	11	7	1	4	8	2	1	8	6	4	10
	3	12	10	2	5	10	3	2	10	10	7	11
	7		11	8	7	11	6	10	12	12	8	
			12		8		7				9	

Formulate the problem of finding a maximal matching as a linear programming problem, and give a printout of your formulation. Use your linear programming package to solve the problem. Write out the schedule implied by your solution and give the number of different programmes that Bhavni can watch.

[MEI, part]

15 Four warehouses, W, X, Y and Z, are to supply four shops, A, B, C and D, with lorry loads of goods. The availabilities of goods at warehouses, the demands for goods at shops, and the costs of sending lorries from warehouses to shops are all shown in the tables below.

Warehouse	W	X	Y	Z
Lorry loads available	5	5	5	5

Shop	A	B	C	D
Lorry loads required	3	7	5	5

Costs	A	B	C	D
W	9	5	1	2
X	4	5	1	3
Y	7	6	2	5
Z	4	7	3	2

(i) The shops are to be supplied from the warehouses at least cost. Formulate this as a linear programming problem, using WA to represent the number of lorry loads sent from warehouse W to shop A, and similarly for WB, etc. Use your linear programming package to solve the problem, and interpret the solution.

Shop D is now to be closed, and as a consequence one warehouse is to be closed. The problem is to find which to close, and how best to supply the remaining shops from the remaining warehouses.

(ii) In a formulation of this problem, IW is an indicator variable which takes the value 0 if warehouse W is closed, and the value 1 otherwise. Explain the purpose of the constraint WA + WB + WC − 5IW = 0, which appears in this formulation.

(iii) Produce a complete formulation of the problem. Use your linear programming package to solve the problem, and interpret the solution.

[MEI]

16 The diagram represents the pipelines forming a gas distribution network. There are three gas wells, X, Y and Z supplying the network, which serves two cities, A and B. The capacity of each pipeline is shown in units of million m^3 per day. Valves are installed in each pipeline so that gas can flow only in the indicated direction.

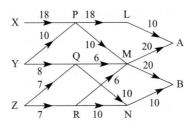

(i) The availabilities of gas at X, Y and Z are 20, 20 and 15 million m³ per day respectively. The requirements for gas at A and B are for 25 and 25 million m³ per day respectively. Explain how to use a super-source, S, and a super-sink, T, to incorporate this information into the network.

(ii) Give the capacity of the cut SXYZPQRLN │ MABT.

(iii) Use the flow-augmenting path SZRNQMBT to augment the flows shown in the diagram below.

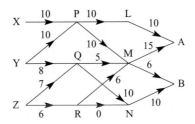

Give the potential capacities on the arcs of the path before augmentation. On a copy of the network show the flows through the network after the augmentation.

(iv) Prove that the total flow after your augmentation in part (iii) is maximal.

(v) Formulate a linear programming problem, the solution of which gives a flow pattern achieving a maximum flow through the network. Produce a printout of your formulation. Use your linear programming software to solve the problem, and give a printout of the solution.

It is proposed to improve the network by laying a new pipeline connecting L and M. This is to have a capacity of 10 million m³ per day, and will allow flows in either direction.

(vi) Modify your formulation to incorporate the new pipeline. Produce printouts of your modified formulation and the new solution.

[**MEI**]

17 The diagram represents a directed network of connected pipes together with weights representing their capacities.

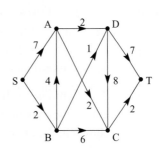

(i) The maximum flow from S to T is 5 units. Give a cut with this capacity.

A flow of 2 units is established along SBADT and a flow of 2 units along SACT.

(ii) Label these flows on a copy of the network, together with potential flows and potential backflows.

(iii) Give a single flow-augmenting path which will achieve a total flow of 5 units. Mark the resulting flow along each pipe on a separate copy of the network.

(iv) Construct a linear programming model to find the maximum flow through the network, using variables such as SA to represent the flow along the pipe from vertex S to vertex A. Use your linear programming package to solve the problem and include a copy of the printout.

(v) The network also models another problem in which the weights now represent distances in the indicated directions. Change your LP formulation so that it finds the shortest distance from S to T. Run your LP, include a copy of the printout, and interpret the solution.

(vi) The arc AD is now changed to be undirected with a weight of 2. Why does this not affect your answer to part (iv)? Change your LP in part (v) and use it to find the new shortest route from S to T.

[MEI]

18 A theatrical agent has to find work for six acts on Saturday. There are two rock bands, two comedians, a solo singer and a girl band.

The Apollo Club (A) wants a rock band or a comedian or a girl band.
Baileys (B) will take a rock band, a comedian or the solo singer.
The Comedy Cave (C) is only suitable for a comedian.
Disco Beat (D) will not take a comedian.
The Ersatz Club (E) will not employ a rock band or a comedian.
The Folio Hotel (F) needs a solo singer.

(i) Represent this information on a bipartite graph, ensuring that you have a separate node for each act.

Provisional bookings exist for a rock band at the Apollo, a rock band at Baileys, a comedian at the Comedy Cave, the solo singer at the Folio Hotel and the girl band at Disco Beat.

(ii) Represent the provisional bookings on a bipartite graph.

(iii) Use a tree search to find an alternating path from the unbooked comedian to the Ersatz Club, and hence find bookings for all of the acts.

(iv) Ignoring the provisional bookings, construct an LP to maximise the number of bookings on Saturday. Produce a printout of your formulation, run the program, and interpret the results.

The agent also wishes to maximise the fees paid for the acts so that she can collect the maximum commission. The fees that the venues will pay are shown in the table on the next table.

Fees payable (£)	A	B	C	D	E	F
Rock band	400	450		300		
Comedian	250	200	150			
Solo singer		200		150	200	250
Girl band	350			400	300	

(v) Adapt your LP to find the booking pattern which will maximise the fees paid. Produce a printout of your formulation, run the LP, and interpret the results.

[MEI]

19 Tasks 1, 2, 3, 4 and 5 are to be undertaken by individuals A, B, C, D and E. Each person has to do one task. Each task has to be undertaken by one person. The table below shows who can do which task.

	A	B	C	D	E
1	✓		✓		
2			✓	✓	
3	✓				✓
4		✓		✓	✓
5		✓		✓	

(i) Formulate as an LP the problem of allocating as many tasks to people as is possible.
(ii) Solve your LP problem, and interpret the solution. Produce printouts of your input and your output.

The costs of employing each individual on each task are now to be taken into account. These costs are shown in the table below.

	A	B	C	D	E
1	20		12		
2			14	30	
3	25				17
4		15		20	18
5		30		20	

(iii) Formulate the problem of finding a minimum cost allocation as an LP.
(iv) Solve your LP problem, and interpret the solution. Produce printouts of your input and your output.

Each task is now to be performed three times. A is contracted to do two tasks, with the same restrictions and costs as above. Similarly B is contracted to do one task, C to do three tasks, D to do five tasks and E to do four tasks.

(v) Formulate an LP to find a minimum cost allocation of tasks to people.

(vi) Solve your LP problem, and interpret the solution. Produce printouts of your input and your output.

[MEI]

KEY POINTS

1 Many practical problems can be modelled as LP problems. These include:

- shortest path problems
- network flow problems
- matching problems
- allocation problems
- transportation problems
- critical path analysis
- maximin problems
- minimax problems
- problems involving indicator variables
- set covering problems
- integer programming problems such as the knapsack problem.

2 Post-optimal information includes:

- the value of each variable which is in the solution
- the reduced cost of each variable which is not in the solution
- the dual price corresponding to each tight constraint
- the slack (or surplus) corresponding to each constraint which is not tight
- ranges of values of the objective coefficients for which the basis will not change
- ranges of values of constraint right-hand sides for which the basis will not change.

7 Recurrence relations

Great fleas have little fleas upon their backs to bite 'em,
And little fleas have lesser fleas and so on ad infinitum.

Augustus de Morgan

ACTIVITY

Work through this algorithm, starting with the word 'mathematics'.

Step 1 Think of an English word and write it down.

Step 2 Write, as a word, the number of letters in the word you have just written.

Step 3 If the last two words you have written down are the same, stop.

Step 4 Go back to step 2.

Repeat this for several different words. Do you always get the same answer? What happens if you do it in French? Or in other languages like Spanish or Welsh?

Iteration

In the activity above you will have produced a sequence of answers, each of which became the starting point for the next run through.

mathematics → eleven → six → three → …

A single run through such an algorithm is called an *iteration*.

In this case the algorithm is given as a sequence of numbered instructions, rather like a computer program, except that it is written in English. This is not the only way of writing an iteration, or indeed the most common. In mathematical examples the algorithm for an iteration is often written in the form of a *recurrence relation*, the rule for going from one term to the next. You also need to know the first term so you can get started.

Here is an example of an iteration given as a recurrence relation.

$$x_{n+1} = \frac{x_n^2 + 9}{2x_n} \text{ with starting value } x_0 = 1$$

Working through this, the algorithm proceeds as follows, to calculator accuracy.

$$x_0 = 1$$

$$x_1 = \frac{1^2 + 9}{2} = 5$$

$$x_2 = \frac{5^2 + 9}{2 \times 5} = 3.4$$

$$x_3 = \frac{3.4^2 + 9}{2 \times 3.4} = 3.023\,529\,412$$

$$x_4 = \frac{3.023\,529\,412^2 + 9}{2 \times 3.023\,529\,412} = 3.000\,091\,554$$

$$x_5 = \frac{3.000\,091\,554^2 + 9}{2 \times 3.000\,091\,554} = 3.000\,000\,001$$

$$x_6 = \frac{3.000\,000\,001^2 + 9}{2 \times 3.000\,000\,001} = 3$$

$$x_7 = \frac{3^2 + 9}{2 \times 3} = 3$$

? Work through this algorithm again, this time replacing the number 9 on the top line by 25. What do you think is happening? Does changing the starting value, x_0, make any difference to the final outcome? What happens if you change 9 to a negative number, say –4?

A surprising way of working out π is to use Wallis's product.

$$\pi = 2 \times \frac{4}{3} \times \frac{16}{15} \times \frac{36}{35} \times \frac{64}{63} \times \dots$$

> The terms on the top line are $2^2, 4^2, 6^2, 8^2, \dots$
> So the general term has $(2n)^2$ or $4n^2$ on the top line and $4n^2 - 1$ on the bottom line

You can write the algorithm for this as the iteration

$$x_{n+1} = x_n \times \left(\frac{4n^2}{4n^2 - 1} \right) \qquad x_1 = 2$$

> The first term need not always be x_0

A recurrence relation may involve more than one term, like this one which generates the Fibonacci sequence: 1, 1, 2, 3, 5, 8, 13, …

$$x_{n+2} = x_{n+1} + x_n \qquad x_0 = 1, x_1 = 1$$

The important thing to notice in these two examples is that after each iteration you have a specific value to use as the starting point for the next iteration. This is always the case in an iterative process.

Recursion

Sometimes the calculation using a recurrence relation can be carried out in two different ways. Suppose you wanted to find x_6 in this example:

$$x_{n+1} = (n+1)x_n \qquad x_0 = 1.$$

You could work it from the bottom up as follows.

$$1 \rightarrow 1 \rightarrow 2 \rightarrow 6 \rightarrow 24 \rightarrow 120 \rightarrow 720$$
$$x_0 \quad x_1 \quad x_2 \quad x_3 \quad x_4 \quad x_5 \quad x_6$$

Or you would work it from the top down as follows.

$$x_6 = 6x_5 = 30x_4 = 120x_3 = 360x_2 = 720x_1 = 720x_0 = 720 \text{ (since } x_0 = 1)$$

The answer, of course, is the same both ways.

The first approach is an iteration, since at each stage there is a specific value to use as the starting point for the next calculation. The second approach is called *recursion*. When you use this method, you do not have a specific value to work with until you arrive at a final answer. Until then, it is always defined in terms of the next step. If you are doing this calculation by hand, the difference between the two methods may seem trivial, but if you are writing a computer program the two methods require fundamentally different styles of program, so the difference is crucial.

The essential feature of recursion is that the solution to a problem is described in terms of solutions to easier or smaller versions of the same problem. You will have realised that in the example above you were finding $6! = 6 \times 5 \times 4 \times 3 \times 2 \times 1$. In the recursive method you did this by using

$6! = 6 \times 5!$
$5! = 5 \times 4!$ and so on.

The solution to finding six factorial was given in terms of finding five factorial, a slightly easier version of the same problem, and so on all the way down to zero factorial which you were given to be one.

Here are some non-mathematical examples of recursion, starting with a well-known story.

> It was a dark and stormy night. The wind howled, and rain lashed the rattling window panes. Three children sat huddled over the embers of a dying fire. The youngest one said, 'Tell us a story'. The oldest one replied:

> 'It was a dark and stormy night. The wind howled, and rain lashed the rattling window panes. Three children sat huddled over the embers of a dying fire. The youngest one said, "Tell us a story". The oldest one replied:

> "It was a dark and stormy night ..." and so on.

You can write this as a structured program, in LOGO, called STORY. Notice how the program refers to itself at the end of the second line.

TO STORY
PRINT "It was a dark ... The oldest one replied:" STORY END.

Note

This is not a very good program because it goes on for ever, calling itself up time and time again. Like most recurrence relations, it should have a stopping condition built in since otherwise it never leads to an 'answer'. The same is true of the next example.

A dictionary definition: Recursion *see recursion*.

First-order recurrence relations

A bank pays interest at 5% per annum on deposit accounts. In one account a deposit of £1000 is left to grow over a number of years.

To model this account mathematically use a recurrence relation in which the time period is one year and in which you start counting when the first deposit is made at time 0. (Sometimes it makes more sense to start at 1 rather than at 0. The resulting formulae then look slightly different.)

A recurrence relation to model the account is

$$u_n = 1.05u_{n-1} \qquad\qquad u_0 = 1000$$

The sum that was previously in the account   The interest which is added on

In a second account annual deposits of £1000 are made.

A recurrence relation to model this second account is

$$u_n = 1.05u_{n-1} + 1000 \qquad\qquad u_0 = 1000$$

Both of these are first-order linear recurrence relations because u_n is expressed as a linear function of its immediate predecessor, u_{n-1} only. If u_{n-2} had also been involved, but no earlier terms, then the recurrence relation would have been of second order.

The first recurrence relation consists entirely of u terms. It is called *homogeneous*. The second has a term not involving u. It is called *non-homogeneous*.

In each case you would like to know how much money is in the account after n years, for whatever value you choose for n. That is you would like to have an explicit expression for u_n, *not* involving u_{n-1} or earlier terms. Such an expression is called a solution of the recurrence relation.

First-order, linear, homogeneous recurrence relations

If $\quad u_n = au_{n-1}$

then $\quad u_n = a(au_{n-2}) = a^2u_{n-2} \ (\text{since } u_{n-1} = au_{n-2})$
$\qquad\quad u_n = a^2(au_{n-3}) = a^3u_{n-3}$
$\qquad\quad\vdots$

$\qquad u_n = a^nu_0 \quad\longleftarrow\quad$ This is the solution

So for £1000, deposited and left, after n years it has grown to be worth $1.05^n \times £1000$.

EXAMPLE 7.1

How long will it take £1000, deposited at 5%, to double in value?

SOLUTION

You need n such that

$$1000 \times 1.05^n \geqslant 2000.$$
$$1.05^n \geqslant 2$$
$$n\log 1.05 \geqslant \log 2$$
$$\text{giving } n \geqslant \frac{\log 2}{\log 1.05} = 14.21$$

> If you know about logarithms, then you can use logs to any base. If you don't know about logs, then you can solve the inequality by iteration

After 14 years £1000 will have grown to £1979.93.

After 15 years it will have grown to £2078.93, so it takes 15 years to double in value.

First-order, linear, non-homogeneous recurrence relations with a constant term

If $\quad u_{n+1} = au_n + b \qquad\qquad a \neq 1$

so that $\quad u_n = au_{n-1} + b$

then $\quad u_n = a^2 u_{n-2} + ab + b$

Repeating the process

$$u_n = a^2(au_{n-3} + b) + ab + b$$
$$= a^3 u_{n-3} + a^2 b + ab + b$$

Eventually

$$u_n = a^n u_0 + a^{n-1}b + a^{n-2}b + \ldots + ab + b$$
$$= a^n u_0 + b(a^{n-1} + a^{n-2} + \ldots + a + 1)$$

The term in brackets is the sum of the first n terms of a geometric progression with first term 1 and with common ratio a. It can be written as $\dfrac{1 - a^n}{1 - a}$ or $\dfrac{a^n - 1}{a - 1}$. (The first is more convenient if $a < 1$ and the second if $a > 1$, but they are algebraically equivalent.)

Thus $\quad u_n = a^n u_0 + \dfrac{b(a^n - 1)}{a - 1} \qquad a \neq 1$

EXAMPLE 7.2

If £1000 is deposited each year in an account paying 5% per annum, after how many years will the accumulated amount exceed £20 000?

SOLUTION

You need n such that

$$(1.05^n \times 1000) + 1000 \left(\frac{1.05^n - 1}{1.05 - 1} \right) \geqslant 20\,000$$

$$\Rightarrow \qquad 1.05^n + \frac{1.05^n - 1}{1.05 - 1} \geqslant 20$$

$$\Rightarrow \qquad 1.05^n + \frac{1}{0.05}\,(1.05^n - 1) \geqslant 20$$

$$\Rightarrow \qquad 21(1.05^n) - 20 \geqslant 20$$

$$\Rightarrow \qquad (1.05^n) \geqslant \frac{40}{21}$$

$$\Rightarrow \qquad n \geqslant \frac{\log(\frac{40}{21})}{\log(1.05)} \approx 13.21$$

After 13 years the sum in the account will be

$$(1.05^{13} \times 1000) + 1000 \left(\frac{1.05^{13} - 1}{1.05 - 1} \right) = £19\,598.63$$

After 14 years the sum in the account will be

$$(1.05^{14} \times 1000) + 1000\, \frac{1.05^{14} - 1}{1.05 - 1} = £21\,578.56$$

First-order, linear, non-homogeneous recurrence relations with a constant term – special case ($a = 1$)

If $\qquad u_{n+1} = u_n + b$

so that $\quad u_n = u_{n-1} + b$

then $\qquad u_n = (u_{n-2} + b) + b = u_{n-2} + 2b$

and $\qquad u_n = (u_{n-3} + b) + 2b = u_{n-3} + 3b$

so $\qquad u_n = u_0 + nb$

Other first-order, linear, non-homogeneous recurrence relations

It is not unusual to need a model of the form $u_{n+1} = ku_n + f(n)$, where the final term, $f(n)$, is more complex than a constant. This makes the result non-linear and, depending on the form of $f(n)$, it can be difficult to handle.

As an example, consider $u_{n+1} = u_n + n \qquad u_0 = 0$.

This is the recurrence relation which describes the construction of the triangle numbers. You can try the same step-by-step approach.

$$u_n = u_{n-1} + n$$

so $\qquad u_n = (u_{n-2} + n - 1) + n = u_{n-2} + (n-1) + n$

and $\qquad u_n = (u_{n-3} + n - 2) + (n-1) + n = u_{n-3} + (n-2) + (n-1) + n$

giving $\quad u_n = u_0 + 1 + 2 + \ldots + (n-2) + (n-1) + n$

$$= 0 + \frac{n(n+1)}{2}$$

An alternative approach

The first-order relations dealt with so far are quite easy to tackle using the step-by-step approach. The hardest was $u_{n+1} = au_n + b \ (a \neq 1)$, for which the solution was $u_n = a^n u_0 + \dfrac{b(a^n - 1)}{a - 1} \left(\text{or } u_n = a^n u_0 + \dfrac{b(1 - a^n)}{1 - a} \right)$.

As an alternative to memorising that result, or to going through the process of deriving it, you can just remember that it is a straight line function of a^n, i.e.

$$u_n = \lambda a^n + \mu \qquad \qquad \qquad ①$$

To find λ and μ first put $n = 0$

giving $\qquad \lambda + \mu = u_0 \qquad \qquad \qquad ②$

then put $\qquad \quad n = 1$

giving $\quad a\lambda + \mu = au_0 + b \qquad \qquad \qquad ③$

Equation ③ − ② gives

$$(a - 1)\lambda = (a - 1)u_0 + b \quad \Rightarrow \quad \lambda = u_0 + \frac{b}{a - 1}$$

Multiplying equation ② by a and subtracting equation ③ gives

$$(a - 1)\mu = -b \qquad \qquad \Rightarrow \quad \mu = \frac{-b}{a - 1}.$$

Substituting for λ and μ in equation ① gives

$$u_n = \left(u_0 + \frac{b}{a - 1} \right) a^n - \frac{b}{a - 1}$$

$$= a^n u_0 + \frac{a^n b}{a - 1} - \frac{b}{a - 1}$$

$$= a^n u_0 + \frac{b(a^n - 1)}{a - 1}$$

so you end with the same result. This process looks easier when you have specific numbers for a and b. Try it on question 1 parts (iii) and (iv) below.

1 For each of the following
   (a) write down the first five terms of the sequence as functions of the initial term
   (b) write the general term as a function of the initial term.

   (i) $u_{n+1} = u_n + 3$ $\qquad \qquad n \geqslant 0$
   (ii) $u_{n+1} = u_n + 3$ $\qquad \qquad n \geqslant 4$
   (iii) $u_{n+1} = 4u_n + 3$ $\qquad \qquad n \geqslant 0$
   (iv) $u_{n+1} = 4u_n + 3$ $\qquad \qquad n \geqslant 4$

2 Calculate the equal annual repayments which are required to pay off a £10 000 loan over 10 years with interest at 8% per annum.

3 Model the sequence 7, 17, 37, 77, 157, ... with a recurrence relation, and solve it to give an expression for the nth term in terms of n.

4 A cup of coffee is at a temperature of 90 °C at time 0. Room temperature is 20 °C, and every second the coffee's temperature drops by $\frac{1}{500}$ of the excess of its temperature above room temperature. Use a spreadsheet to find how long it takes for the coffee to cool to 70 °C.

5 In the 'Tower of Hanoi' the object is to transfer a pile of rings from one needle to another, one ring at a time, in as few moves as possible, with never a larger ring sitting upon a smaller ring.

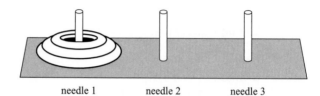

needle 1   needle 2   needle 3

In the Temple of Benares, each day a priest moves a ring from an original pile of 64 according to the Tower of Hanoi rules. A novice priest asks for instruction on how to continue with the task. He is told:

'To move one ring from needle 1 to needle 2 or 3, move it directly.'
'To move n rings, where $n > 1$, from needle 1 to needle 3, first move $n - 1$ rings from needle 1 to needle 2, then move the nth ring from needle 1 to needle 3, then move the $n - 1$ rings from needle 2 to needle 3.'

(i) Apply the recursive procedure defined by the instructions to move three rings from needle 1 to needle 3. Show each move.

(ii) Calculate how many days it will take for the priests of Benares to complete their task.

6 Ian operates his bank account according to strict rules. Each month he puts £400 into the account and allows himself to spend one fifth of what was in the account at the end of the previous month.

(i) Letting u_n be the amount in the account at the end of month n, write down a recurrence relation for u_{n+1} in terms of u_n.

(ii) Find a general expression for u_n, given that $u_1 = 400$.

(iii) Starting with an empty account, how much will Ian have in his account,

(a) after 12 months

(b) in the long run?

[Oxford]

Second-order recurrence relations

Rabbits breed quickly! Suppose that you start with one newborn pair and that every month any mature pair gives birth to a new pair. It takes two months for a newborn pair to mature.

The progression of the population is shown in the table below.

Time (months)	0	1	2	3	4	5	6	...
Number of newborn pairs	1	0	1	1	2	3	5	...
Number of 1-month-old pairs	0	1	0	1	1	2	3	
Number of mature pairs ($\geqslant 2$ months old)	0	0	1	1	2	3	5	...
Total number of pairs (u_n)	1	1	2	3	5	8	13	...

The sequence, u_n, is attributed to Fibonacci (*c.* 1170–1250).

The recurrence relation which describes the sequence is

$$u_{n+2} = u_{n+1} + u_n \qquad u_0 = 1, u_1 = 1$$

This is a second-order recurrence relation. Note that it requires *two* initial conditions.

Historical note

Fibonacci was a nickname for Leonardo Pisano. He was probably born in Pisa, Italy, in 1170. His father was a diplomat and he grew up in North Africa, where mathematics was much more advanced than in Europe. When Fibonacci returned to Pisa in the year 1200, he wrote a number of books on mathematics, all by hand since printing had not then been invented.

Some have been lost but, amazingly, others survive. While his work did include original discoveries in number theory, his long-term contribution was probably much greater in exposing Europe to the sophistication and learning of Africa and Asia, including the Indian–Arabic system for writing numbers that we use now in place of the Roman system.

Second-order, linear, homogeneous recurrence relations

A recurrence relation of this type has the form

$$u_{n+2} = au_{n+1} + bu_n$$

Try the solution $u_n = Am^n$, where A and m are constants. This would give

$$Am^{n+2} = aAm^{n+1} + bAm^n$$

This collects and factorises to

$$Am^n(m^2 - am - b) = 0.$$

Since you cannot have $A = 0$ or $m = 0$, it must be that

$$m^2 - am - b = 0.$$

This is known as the *auxiliary equation*. It is a quadratic equation with two, possibly complex, solutions. Let the solutions be m_1 and m_2.

There are three possibilities to consider:

- $m_1 \neq m_2$ and both are real
- $m_1 = m_2$ (in which case it is real)
- $m_1 \neq m_2$ and both are complex.

$m_1 \neq m_2$ and both are real

In this case both $u_n = m_1{}^n$ and $m_2{}^n$ are solutions.

It can be shown that any linear combination (i.e. a sum of multiples) of the two solutions is also a solution. Furthermore it can be shown that *any* solution of the recurrence relation is of this form, i.e.

$$u_n = Am_1{}^n + Bm_2{}^n.$$

EXAMPLE 7.3

Find the specific solution of

$$u_{n+2} = u_{n+1} + 6u_n \qquad u_0 = 0, \ u_1 = 1.$$

SOLUTION

The auxiliary equation is

$$m^2 - m - 6 = 0$$
$$\Rightarrow \quad (m-3)(m+2) = 0$$
$$\text{so} \qquad\qquad m = 3 \text{ or } -2.$$

The general solution is

$$u_n = A3^n + B(-2)^n.$$
$$u_0 = 0 \Rightarrow \qquad 0 = A + B$$
$$u_1 = 1 \Rightarrow \qquad 1 = 3A - 2B$$

> This is the general solution. To find the specific solution with the given initial conditions those conditions are substituted to give two simultaneous equations for A and B

Solving gives $A = \frac{1}{5}$, $B = -\frac{1}{5}$.

So the specific solution is

$$u_n = \tfrac{1}{5} 3^n - \tfrac{1}{5}(-2)^n.$$

Check:

n	0	1	2	3	4	5
u_n	0	1	1	7	13	55

e.g. $\quad u_5 = \tfrac{1}{5} 3^5 - \tfrac{1}{5}(-2)^5$
$$= \tfrac{1}{5}(243 - (-32))$$
$$= \tfrac{1}{5} \times 275 = 55$$

$m_1 = m_2$ (= m, say)

In this special case it can be shown that the general solution is

$$u_n = Am^n + Bnm^n.$$

EXAMPLE 7.4

Find the specific solution of

$$u_{n+2} = 8u_{n+1} - 16u_n \qquad u_0 = 1, u_1 = 1.$$

SOLUTION

The auxiliary equation is

$$m^2 - 8m + 16 = 0$$
$$\Rightarrow \qquad (m-4)^2 = 0$$
$$\text{so} \qquad m = 4.$$

The general solution is

$$u_n = A4^n + Bn4^n$$
$$u_0 = 1 \Rightarrow \qquad 1 = A$$
$$u_1 = 1 \Rightarrow \qquad 1 = 4A + 4B$$

Solving gives $A = 1$, $B = -\frac{3}{4}$.

So the specific solution is

$$u_n = 4^n - \frac{3}{4}n4^n$$
$$= (1 - \frac{3n}{4})4^n.$$

Check:

n	0	1	2	3	4	5
u_n	1	1	−8	−80	−512	−2816

e.g. $\qquad u_5 = (1 - \frac{15}{4})4^5$
$$= -2816$$

$m_1 \neq m_2$ and both are complex

In this case the solution looks similar to the real case, i.e. $u_n = Am_1{}^n$ and $Bm_2{}^n$.

Although m_1 and m_2 are complex numbers, the expression produces a real answer. Complex number theory allows the general solution to be written in terms of sines and cosines, showing it to be cyclic in nature.

You are not required to be able to produce an analytical (i.e. theoretical) solution in this case.

Use a spreadsheet to investigate the behaviour of

$$u_{n+2} = u_{n+1} - 2u_n$$

for various values of u_0 and u_1.

Second-order, linear, non-homogeneous recurrence relations

A recurrence relation of this type has the form

$$u_{n+2} = au_{n+1} + bu_n + f(n).$$

It can be shown that the general solution is

$$u_n = \left(\begin{array}{c}\text{general solution to the associated}\\ \text{homogeneous equation}\end{array}\right) + \left(\begin{array}{c}\text{one particular solution}\\ \text{to the full equation}\end{array}\right)$$

For example, look at

$$u_{n+2} = u_{n+1} + 6u_n + (n + 2) \qquad u_0 = 1, u_1 = 1.$$

You know from Example 7.3 that the general solution to the homogeneous equation $u_{n+2} = u_{n+1} + 6u_n$ is

$$u_n = A3^n + B(-2)^n.$$

You now need a particular solution to the full equation. Try $u_n = pn + q$ so that $u_{n+1} = p(n + 1) + q$ and $u_{n+2} = p(n + 2) + q$.

For this to work you need (substituting these expressions into the recurrence relation)

$$p(n + 2) + q = p(n + 1) + q + 6(pn + q) + n + 2$$
$$pn + (2p + q) = (7p + 1)n + (p + 7q + 2)$$

So you will need (looking at the coefficients of n)

$$p = 7p + 1, \text{ giving } p = -\tfrac{1}{6}$$

and (looking at the constant terms)

$$2p + q = p + 7q + 2, \text{ giving } q = \frac{p - 2}{6} = -\tfrac{13}{36}.$$

You now have the general solution

$$u_n = A3^n + B(-2)^n - \tfrac{1}{6}n - \tfrac{13}{36}.$$

Finally, use the initial conditions to find the specific solution for the given initial conditions.

$$u_0 = 1 \Rightarrow \quad 1 = A + B - \tfrac{13}{16}$$
$$\text{so} \qquad A + B = \tfrac{49}{36}$$
$$u_1 = 1 \Rightarrow \quad 1 = 3A - 2B - \tfrac{1}{6} - \tfrac{13}{36}$$
$$\text{so} \qquad 3A - 2B = \tfrac{55}{36}.$$

Solving gives $A = \frac{17}{20}, \ B = \frac{23}{45}.$

So the specific solution is

$$u_n = \tfrac{17}{20} 3^n + \tfrac{23}{45}(-2)^n - \tfrac{n}{6} - \tfrac{13}{36}.$$

Check:

n	0	1	2	3	4	5
u_n	1	1	9	18	76	189

e.g. $u_5 = \tfrac{17}{20} 3^5 + \tfrac{23}{45}(-2)^5 - \tfrac{5}{6} - \tfrac{13}{36}$
$= 189.$

Clearly this approach hinges on your ability to spot an appropriate particular solution to the full equation. The table below shows some.

$f(n)$	Form of particular solutions
constant	p
n	$pn + q$
n^2	$pn^2 + qn + r$
k^n	pk^n (or pnk^n in special cases)

1 Find the general solutions of

(i) $u_{n+2} = 4u_{n+1} - 4u_n$

(ii) $u_{n+2} - 6u_{n+1} + 9u_n = 0.$

2 Find the specific solution of

$$u_{n+2} - 6u_{n+1} + 9u_n = 0 \qquad\qquad \text{when } u_1 = 1 \text{ and } u_2 = 5.$$

3 Solve the recurrence relation $u_{n+2} = u_{n+1} + 12u_n$ with $u_0 = 1$ and $u_1 = 3.$

Implement a solution on a spreadsheet.

4 Find an expression for the nth Fibonacci number.

5 Paving slabs of dimensions $1 \text{ m} \times 0.5 \text{ m}$ are to be laid to make a path which is 1 m wide. Show that the recurrence relation $u_{n+2} = u_{n+1} + u_n$ models the number of possible ways of laying such a path.

6 Find the specific solution of

$$u_{n+1} - 5u_n + 6u_{n-1} = 1 + (n + 1)^2 \quad \text{when } u_0 = 2 \text{ and } u_1 = 3.$$

Note

You might find the following rule helps to summarise how to deal with special cases. When the solution would otherwise have two terms of the same kind, multiply one of them by n before finding the constants. For example, this happens if an auxiliary equation has equal roots, or if the particular solution is of the same form as a term of the solution of the homogeneous equation.

EXERCISE 7C

Wherever possible, spreadsheet models should be constructed to confirm theoretical results.

1 (i) Solve the recurrence relation $u_{n+1} = au_n + b$ (where a and b are constants, $u_0 = c$ and $a \neq 1$).

   (ii) Sketch a graph of u_n against n in the case where $a = 0.5$ and $b = 10$.

   [AEB, adapted]

2 (i) (a) Solve the recurrence relation $u_{n+2} = 12u_n - u_{n+1}$, given that $u_0 = 0$. and $u_1 = 2$.

   (b) Find u_4 by repeatedly applying the recurrence relation in part (i)(a).

   (c) Find u_4 by applying your result from part (i)(a).

   (ii) (a) Show that the sequence $-0.2, -0.2, -0.2, \ldots$ satisfies the recurrence relation $v_{n+2} = 12v_n - v_{n+1} + 2$, and hence that $v_n = -0.2$ is a particular solution.

   (b) Hence find the complete solution to the recurrence relation in part (ii)(a), given that $v_0 = 0$ and $v_1 = 2$.

   [AEB]

3 The cooling system of my car has a capacity of 10 litres. It contains antifreeze at a concentration of 0.25, so there are 2.5 litres of ethylene glycol and 7.5 litres of water. The system leaks 0.1 litres per week.

   I wish to top it up weekly with 0.1 litres of undiluted ethylene glycol, until the concentration reaches 0.5. Thereafter I shall use a diluted top-up (concentration 0.5) to maintain the concentration.

   (i) Derive a recurrence relation for v_k, the volume of ethylene glycol in the cooling system immediately after the kth top-up. Hence show that the recurrence relation for the concentration of ethylene glycol is $c_{k+1} = 0.99c_k + 0.01$, where c_k is the concentration immediately after the kth top-up.

   (ii) Solve the recurrence relation for c_k. Find the concentration after four weeks and find how many weeks are required for the concentration to reach 0.5.

4 An engine's cooling system has a capacity of 10 litres. It leaks at the rate of 1 litre per week. Initially, at time 0, the system is full, and contains water only. Every week (i.e. at times 1, 2, 3, …) it is topped up with a litre of antifreeze mixture of concentration 25% (a mixture of 750 ml of water and 250 ml of antifreeze).

(i) What is the concentration of antifreeze in the cooling system after it has been topped up at time 1?

(ii) How much antifreeze will leak from the system during the following week?

(iii) How much antifreeze is there in the system after it has been topped up at time 2?

What is its concentration?

Let u_n be the number of millilitres of antifreeze in the system immediately after the top-up at time n. Let c_n be the concentration.

(iv) Express u_{n+1} in terms of u_n.

(v) Express c_{n+1} in terms of c_n.

(vi) Give the long-run concentration, i.e. the value to which c_n tends as n becomes large.

[AEB]

5 Sabina is considering taking out a loan of £30 000 to purchase a flat. She will pay interest each month on the outstanding balance at a rate of 0.95% per month. She will make monthly repayments.

Each month, interest will be added, and then her payment will be used to reduce the outstanding balance.

Let her monthly payment be £X and the amount outstanding after n months be £u_n. Then the recurrence relation giving the amount outstanding after $n + 1$ months is $u_{n+1} = 1.0095u_n - X$.

(i) Explain the structure of this recurrence relation. Explain the construction and the purpose of the 1.0095.

(ii) Solve the recurrence relation to obtain an expression for the amount outstanding after n months as a function of n and X.

(iii) Find the value of X required to pay off the loan in 25 years.

[Oxford, adapted]

6 Victoria has an income of £1000 per month, which goes into her bank account on the first day of each month. On the first day of each month she withdraws spending money from the account. Also at the beginning of each month interest of 1% of the amount in the account during the previous month is added into the account. Throughout the rest of each month the amount in her account remains constant.

After the £1000 deposit, withdrawal of spending money and payment of interest on 1st January, she had a balance of £2000 in the account.

Victoria decided that she would withdraw spending money of £500 on 1st February. Thereafter on the first of each month she would withdraw 25% of the amount in her account (after income, withdrawal and interest) two months previously. Thus on 1st March she would again transfer £500, because 25% of £2000 is £500, but on 1st April she would transfer 25% of the balance (after income, withdrawal and interest) on 1st February.

(All withdrawals and interest payments are computed to the nearest penny.)

(i) (a) Find how much she has in her account during February.

   (b) Find how much she has in her account during March.

(ii) (a) Show that she withdraws £630 on 1st April.

   (b) Find how much she withdraws on 1st May.

(iii) Counting January as month 1, if u_n is the balance in her account during month n, explain why for $n \geqslant 1$,

$$u_{n+2} = 1.01u_{n+1} - 0.25u_n + 1000.$$

(iv) By setting $u_{n+2} = u_{n+1} = u_n$, find the long-term balance in Victoria's account. Say how much she withdraws to spend each month in the long term.

(v) If Victoria were to decide to transfer 50% of the balance instead of 25% then her balances for the first three months would be £2000, £2520 and £2545.20. Say what the amount in her account would be in the long term. Say how much she would withdraw to spend each month in the long term.

[AEB]

7 A maths teacher intends to set members of her class a problem involving matchsticks. As the first stage she wishes to demonstrate the construction of complete $n \times n$ lattices as shown in the diagram.

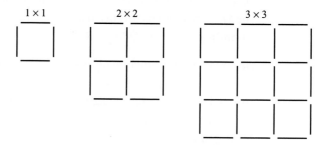

(i) If u_n is the number of matchsticks needed to construct an $n \times n$ lattice, explain why another $4n + 4$ matchsticks are needed to extend it to an $(n + 1) \times (n + 1)$ lattice. Express this in the form of a recurrence relation.

(ii) Solve your recurrence relation to obtain an expression for the number of matchsticks needed to construct an $n \times n$ lattice.

(iii) How many matchsticks will the teacher need to construct a 6×6 lattice?

(iv) The teacher wishes to construct a complete series of lattices from 1×1, through 2×2, 3×3, etc., up to $n \times n$. She wishes to find an expression for s_n, the number of matchsticks that she will need. Find a recurrence relation for s_n.

[AEB, adapted]

8 A fishing club's waters are stocked with a particular variety of fish. Each season breeding increases the population of fish by a *proportion*, b, of the population at the start of the season. Similarly a *proportion*, d, of the population at the start of the season die during the season. Also a *number*, c, of fish are caught and removed from the waters each season.

(i) Write down a recurrence relation in the form $u_{n+1} = pu_n + q$, where u_n is the fish population at the start of the nth season and p and q are constants to be expressed in terms of b, c, and d.

(ii) Solve the recurrence relation in part (i) to find a general expression for u_n in terms of p, q and u_1.

(iii) Given that $p = 1.08$ and that $q = -160$, find the size of the fish population which would remain constant from season to season.

(iv) The club is anxious not to have to restock by buying in fish each year since this is expensive. They propose to achieve this by ensuring that the initial stock is above the equilibrium population. What does the mathematical model which you have developed predict will happen under these circumstances? What is actually likely to happen and why?

[AEB]

9 A plane is divided by straight lines.
  (i) One line divides the plane into two regions.
     Two lines divide the plane into four regions.
     Three lines divide the plane into a maximum of seven regions.
     (a) Continue this sequence for four, five and six lines.
     (b) Establish a recurrence relation for u_{n+1}, the maximum number of regions created by $n + 1$ lines, in terms of u_n. Justify your recurrence relation by a geometrical argument.
     (c) Solve your recurrence relation for u_n.

  (ii) Two lines intersect in at most one place.
     Three lines have a maximum of three points of intersection.
     Four lines have a maximum of six points of intersection.
     Find and solve a recurrence relation for the maximum number of intersections of n straight lines. Justify your recurrence relation by a geometrical argument.

[Oxford]

10 During a busy period a large computer installation handles messages through a queuing system. In any one-second period the number of messages joining the queue can be modelled by the expression $900 - 1.75Q$, where Q is the number of messages in the queue at the beginning of the period (provided $0 \leqslant Q \leqslant 500$). The expression is a reducing function of Q, modelling the fact that as the queue increases the service deteriorates. This leads to a decrease in the rate at which messages join the queue.

The computer can deal with, on average, 200 messages per second, so that 200 messages leave the queue every second.

  (i) Explain how to produce an expression for Q_{n+1}, the number of messages in the queue at the time $n + 1$ seconds, in terms of Q_n, the number in the queue at time n. Show that $Q_{n+1} = 700 - 0.75Q_n$.

  (ii) Given that there are 500 messages in the queue at time 0, derive an expression in terms of n only for the number of messages in the queue at time n.

(iii) In the long run, how many messages would you normally expect to see in the queue?

(iv) What would be the problem in using a similar approach to model the queue at a newspaper kiosk?

[AEB]

11 A piece of coral has four new branches. During each year one new branch develops on each branch which is more than one year old.

The number of branches after n years is denoted by u_n. Thus $u_0 = 4$, $u_1 = 4$ and $u_2 = 8$.

(i) Find u_3, the number of coral branches after three years.

(ii) Explain why $u_{n+1} = u_n + u_{n-1}$ $\qquad (n \geqslant 1)$.

(iii) (a) Solve the recurrence relation to find an exact expression for u_n in terms of n.

(b) Verify that your expression gives $u_2 = 8$ exactly.

(iv) Construct a spreadsheet model for u_n. Use it and adapt it to answer the following questions.

(a) How many branches are there at the end of year 5 (i.e. u_5) and how many of these are more than one year old (i.e. u_4)?

At the *end* of year 5 a number of old branches are removed (without losing any of the newer branches), so that the total number of branches at the end of year 6 is 34.

(b) How many old branches are removed at the end of year 5?

(c) If no further branch removal takes place, how many branches will there be at the end of year 10, and how many at the end of year 20?

(d) At the end of year 6, how many of the 34 branches are old and how many new? Show that it is not possible to achieve between 120 and 126 branches at the end of year 10 by removing old branches at the end of year 6.

12 The government of Euroland has a policy of supporting agricultural prices by buying surplus produce in times of plenty. This produce is kept in *intervention stores*. At the present time there are 250 000 tonnes of soft grain in the soft intervention store.

Each year birds and vermin consume 5% of the stored grain, another 5% is disposed of as rotten and 10% is donated to famine relief. The current level of supply and demand for soft grain necessitates the government buying in 20 000 tonnes per year.

(i) Assuming this situation to continue, express the above information as a recurrence relation for u_n, where u_n is the quantity of soft grain stored in n years' time.

(ii) How many tonnes should the government expect to have to store in five years' time? How many tonnes should the government expect to have to store in n years' time?

(iii) In how many years will the government be able to reduce soft grain storage capacity to 105 000 tonnes?

(iv) How much storage capacity will be needed in the long run?

(v) Suppose that three years from now the soft grain crop fails. For that year only, instead of buying in, the government has to hand out 100 000 tonnes from the store. Repeat parts (i), (ii) and (iv), taking account of this new information.

Describe how the yearly change in the quantity of stored grain has been altered by the crop failure.

[Oxford]

13 Grundy the gardener cuts the lawn to a length of 1 cm. He notes that, on average, the growth during the $(n + 1)$th day after a cut is 0.8 times the growth during the nth day after a cut. His report on this to Lord Whimsey states that $G_{n+1} = 0.8G_n$.

Lord Whimsey, keen to use his mathematical ability to organise the estate's activities efficiently, decides to model L_n, the length of the grass n days after a cut.

(i) Explain why the recurrence relation of L_{n+1} in terms of L_n and G_n is
$L_{n+1} = L_n + G_{n+1}$.

(ii) Given that $G_{n+1} = 0.8 \times G_n$ and that $G_n = L_n - L_{n-1}$, produce a second-order recurrence relation for L_{n+1} in terms of L_n and L_{n-1}.

(iii) Given that $G_1 = 1$, solve your recurrence relation to find an expression for L_n in terms of n.

Grundy has no time for new-fangled ideas such as second-order recurrence relations. He argues that if $G_{n+1} = 0.8G_n$ and $G_1 = 1$, then $G_n = (0.8)^{n-1}$ and

$$\sum_{i=1}^{n} G_i = 5 - 5(0.8)^n.$$

(iv) Grundy aims to cut the lawn when it reaches a height of about 4 cm. How often should he cut the lawn (to the nearest whole number of days)?

(v) If the lawn grows for 200 days per year, approximately what will be the total length of grass that he cuts in a year?

(vi) What will be the approximate total length cut in the year if he cuts it only when it has reached a length of 4.7 cm?

[AEB]

14 Indiana Smith wants to cross the desert, a 20-day trip. He has a team of 19 porters to help him.

All 20 members of the party will carry supplies on day 1, at the end of which a supply camp will be established and one porter will be sent back. The remaining party of 19 will repeat the exercise on day 2, establishing a supply camp and sending one porter back. The returning porter will use the supplies left in the earlier camps on his return journey.

The plan is for this to be repeated until, on the final day, only Indiana will be left to achieve his objective.

He and his porters each drink one can of water during each day.

Let C_k be the camp which is k days from the destination. Let w_k be the number of cans of water that have to be transported to camp C_k. Thus $w_0 = 0$, since Indiana does not wish to return.

(i) Show that $w_1 = 2$.

(ii) The water transported to camp C_{k+1} has to provide for that required at C_k, plus sufficient for the on-going party to proceed to C_k, plus sufficient for all except Indiana to return to C_{k+2} in due course.
Show that this implies that $w_{k+1} = w_k + 2(k+1)$.

(iii) Given that the solution to the recurrence relation $w_{k+1} = w_k + 2(k+1)$ is of the form $w_n = an^2 + bn + c$, find the solution of the relation with $w_0 = 0$.

(iv) Use your relation to find out how many cans the party must set out carrying.

Indiana now decides that he will make provision for himself to return as well. Let y_k be the number of cans of water that now have to be transported to C_k.

(v) What is y_0? What is y_1?

(vi) Find a recurrence relation connecting y_{k+1} and y_k.

(vii) By solving the recurrence relation or otherwise, find how many cans the party must set out carrying.

[Oxford]

15 (i) Sequences x_n and y_n for $n = 0, 1, 2, \ldots$, are defined by

$$x_{n+1} = 0.4y_n - 9 \quad \text{and} \quad y_{n+1} = 6.4x_n - 36,$$

where $x_0 = 17$ and $y_0 = 60$.

(a) Use the second recurrence relation to write down an expression for y_n in terms of x_{n-1}. Substitute this expression into the first recurrence relation to give a second-order relation for x_{n+1} in terms of x_{n-1}.

(b) Find x_1 and solve your second-order recurrence relation to give a general expression for x_n.

(c) List the values of x_2 to x_{10}.

(ii) A pet shop owner breeds and sells a species of reptile and its eggs. His stock at the end of month n is x_n breeding pairs of reptiles and y_n eggs, where x_n and y_n are modelled by the recurrence relations in part (i).

(a) If the pet shop owner has 17 breeding pairs and 60 eggs in stock at the end of month 0, produce a spreadsheet with columns showing the number of breeding pairs and the number of eggs at the ends of each of the next nine months. Produce a printout of your spreadsheet.

The model is not realistic and is refined to read $x_{n+1} = [0.4y_n - 9]$ and $y_{n+1} = [6.4x_n - 36]$, where $[z]$ means the integer part of z if z is positive, and zero otherwise, e.g. $[1.6] = 1$, $[1] = 1$, $[-0.6] = 0$.

(b) Add new columns to your spreadsheet to show the modifications to the model. Produce printouts of your new columns, one in which the initial number of breeding pairs is 17 and the initial number of eggs is 60, and one in which the initial numbers are 14 and 60 respectively. Find initial numbers which give an equilibrium.

[MEI]

16 The Chancellor of the Exchequer in Mathland wishes to achieve a steady rate of inflation of 5% per annum in the Mathland economy. He has a Budget each autumn in which he puts in place measures which are intended to achieve this. Unfortunately he does not realise that his measures only take effect after two years.

In autumn 1999 the percentage rate of inflation in Mathland was 5.5, and the Chancellor announced measures to reduce the percentage rate of inflation by 0.5. However these measures will not take effect until autumn 2001. In autumn 2000 earlier actions produced a percentage rate of inflation of 4.8. The Chancellor therefore put in place measures to increase it by 0.2, not realising that this increase will not take effect until autumn 2002.

(i) Show that the percentage inflation rate in autumn 2001 will be 4.3.

(ii) Given that the Chancellor continues to pursue the same policy, find the percentage inflation rates for autumn 2002, 2003, 2004, 2005, 2006 and 2007. Describe what will happen subsequently.

Let u_n be the percentage inflation rate in the autumn of year $2000 + n$, so that $u_0 = 4.8$, $u_1 = 4.3$, etc. Then a recurrence relation describing the effect of the Chancellor's policy is

$$u_{n+2} = u_{n+1} + (5 - u_n).$$

(iii) Construct a spreadsheet in which column A contains the numbers $i = 0, 1, 2, 3, \ldots, 40$, and in which column B contains the values of u_i, $i = 0, 1, 2, 3, \ldots, 40$.

A group of the Chancellor's advisors say that his policies will fail because he is not sufficiently bold. They advise him to aim for an adjustment in each Budget which is 25% greater than the difference between the target of 5 and the actual percentage inflation rate. Thus the recurrence relation would be

$$u_{n+2} = u_{n+1} + 1.25(5 - u_n).$$

(iv) Add column C to your spreadsheet showing the percentage inflation rates that would be produced by following this policy for 40 years, again starting with $u_0 = 4.8$ and $u_1 = 4.3$. Negative rates of inflation are never seen in Mathland, and any negative values should be replaced by 0 as they are produced. Show the formula that you used for an entry in column C.

Another group of advisors say that the Chancellor is being too heavy-handed, and that he should aim for an adjustment which is 75% of the difference between the target of 5 and the actual percentage inflation rate.

(v) Add column D to your spreadsheet showing the percentage inflation rates that would be produced by following this policy for 40 years, again starting with $u_0 = 4.8$ and $u_1 = 4.3$. Produce a printout of your spreadsheet, showing columns A, B, C and D.

The Chancellor is so convinced by the second group's argument that he decides to aim for an adjustment which is only 9% of the difference between the target of 5 and the actual percentage inflation rate. Thus the inflation rate follows the recurrence relation $u_{n+2} = u_{n+1} + 0.09(5 - u_n)$, with $u_0 = 4.8$ and $u_1 = 4.3$.

(vi) Solve this recurrence relation to find an expression for u_n in terms of n.
(vii) Describe the consequences of following this policy.

[MEI]

17 A drug therapy involves administering 200 units of a drug to the patient at time $t = 0$. The drug will then be slowly excreted. One day later, at time $t = 1$, a blood sample is taken and sent to the laboratory, so that in another two days' time, at time $t = 3$, it will be known how much drug remained in the patient at time $t = 1$. This is repeated at time $t = 2$ and subsequently.

Starting at time $t = 3$ the results of the first test are used to calculate a top-up dose of the drug. The top-up dose at time $t = n + 2$, if one is required, is the difference between 200 units and the amount of drug in the patient's body at time $t = n$. No top-up dose is given at time $t = n + 2$ if the amount of drug present at time $t = n$ is 200 units or more.

Suppose that during the course of a day patient X excretes 25% of the drug that was in his body at the beginning of the day. Let x_n be the amount of drug in patient X at time $t = n$.

(i) Explain why $x_{n+2} = 0.75x_{n+1} + \max[(200 - x_n), 0]$ for $n = 1, 2, \ldots$, where $x_1 = 150$ and $x_2 = 112.5$.

(ii) Create a spreadsheet in which column A represents time in days, and in which the entries are the numbers 1, 2, ..., 20, 21. Column B should contain the amount of drug in patient X at the corresponding time.

Patient X's doctor is worried about the fluctuating level of the drug that would be created in such a patient by the therapy. She wonders if it would help to administer a top-up dose which is the difference between 250 units and the amount of drug in the patient's body at time $t = n$ (with no top-up if the amount present at time $t = n$ is 250 units or more).

(iii) Investigate the effects of this revised therapy by calculating revised drug levels in column C of your spreadsheet. Briefly describe the effect of the change.

As another alternative the doctor wonders whether it might be worth trying a top-up dose which is half of the difference between 200 units and the amount of drug in the patient's body at time $t = n$, if that amount is less than 200 units.

(iv) Investigate the effects of this therapy by calculating revised drug levels in column D of your spreadsheet. Describe what would happen if this therapy was used.

The laboratory installs new equipment which enables the results to be returned in one day instead of in two days. Thus the therapy is described by the recurrence relation $x_{n+1} = 0.75x_n + (200 - x_n)$ for $n = 1, 2, \ldots$, with $x_1 = 150$.

(v) Solve this recurrence relation to find x_n in terms of n, and describe what happens under the therapy now.

(vi) Use your spreadsheet to compare your results in part (v) to the results obtained by dispensing with the blood tests and administering a constant top-up dose of 40 units, starting at time $t = 1$. Print out your completed spreadsheet.

[MEI]

18 'Shoot the Bunny' is a fairground game in which a target (the 'bunny') moves to the right or left on a screen at one-second intervals. The competitor tries to 'shoot' the target with a laser gun, which can be fired once a second. The centre of the screen is called the bunny's 'burrow'. The bunny is programmed to move each second as follows.

● It moves towards its burrow by an amount which is 20% of its distance away from its burrow.

● In addition to the above movement, it moves away from the point which the last shot hit by an amount which is 15% of its distance from that point.

The laser gun has a built-in time delay of one second. So when the bunny moves, the last shot hits the point where the bunny was two seconds before.

(i) The bunny starts 30 cm to the right of the burrow ($u_0 = 30$). At time $t = 1$ it moves 6 cm to the left ($u_1 = 24$). Between times $t = 1$ and $t = 2$ the first shot hits the point 30 cm to the right of the burrow.

  (a) Show that the bunny's next position is 18.3 cm to the right of the burrow.

  (b) Explain why, for $n \geq 0$, $u_{n+2} - 0.95u_{n+1} + 0.15u_n = 0$.

  (c) Solve the recurrence relation in part (i)(b), given that $u_0 = 30$ and $u_1 = 24$.

  (d) Describe what happens as the competitor continues to shoot, either by considering your solution to the recurrence relation or by using EXCEL to model the relation.

(ii) The game is adjusted so that the 20% is changed to 25%, and the 15% is also changed to 25%.

   (a) Show that the recurrence relation is now $u_{n+2} - u_{n+1} + 0.25u_n = 0$.

   (b) Solve the recurrence relation in part (ii)(a), given that $u_0 = 30$ and $u_1 = 22.5$.

   (c) Compared to your answer to part (i)(d), describe what happens as the competitor continues to shoot, either by considering your solution to the recurrence relation or by using EXCEL to model the relation.

(iii) In a third situation the bunny is programmed to move halfway towards its burrow, and in addition to move away from the point which the last shot hit by an amount equal to half of its distance from that point.

   (a) Give the new recurrence relation.

   (b) Given that $u_0 = 30$ and $u_1 = 15$, use EXCEL to investigate the bunny's subsequent movements as the competitor continues to shoot. Produce a printout of your formulae and your results, and describe what happens.

[MEI]

19 Each clockwise turn of the control valve of Jonathan's shower through an angle of 10° causes an increase of temperature of 1°C, but only after a 10-second delay, while the water already beyond the control valve emerges. Similarly each anti-clockwise turn through an angle of 10° causes a decrease of temperature of 1°C, but only after a 10-second delay.

Jonathan likes to have his shower water at a temperature of 38°C.

(i) The shower water is currently at a temperature of 21°C. Give the sequence of temperatures if Jonathan were to turn the valve clockwise 20° if the water is too cold, anti-clockwise 20° if the water is too hot, wait for the change to take effect, and then repeat.

(ii) In fact Jonathan is impatient. His first strategy is to turn the valve clockwise or anti-clockwise by 20°, as described in part (i), but every 5 seconds. The shower water is currently at a temperature of 21°C. Give the subsequent sequence of temperatures.

Jonathan finds that he cannot achieve a comfortable temperature, and he tries a second strategy. Initially he turns the valve clockwise or anti-clockwise 20°, then after every 5 seconds he turns it clockwise or anti-clockwise as before, but only by half as much as he turned it 5 seconds before.

(iii) Explain how the following recurrence relation models the water temperature at 5-second intervals under this second strategy.

$$u_{n+2} = u_{n+1} - \frac{x_n}{2^{n-1}}$$

where $u_0 = u_1 = 21$ $n = 0, 1, 2, \ldots$

$x_n = 1$ if $u_n > 38$ $x_n = -1$ if $u_n < 38$ $x_n = 0$ if $u_n = 38$

(iv) Build a spreadsheet to show how the water temperature changes over a period of one minute when Jonathan uses this second strategy. Produce a printout of your results and comment on them.

Jonathan is still unhappy with his control strategy. He decides to keep it as before, but to start by turning the valve through 90°.

(v) Explain why the recurrence relation

$$u_{n+2} = u_{n+1} - \frac{9x_n}{2^n}$$

with $u_0 = u_1 = 21$ and with x_n defined as in part (iii), models the water temperature at 5-second intervals when Jonathan uses his third strategy.

(vi) Build a spreadsheet to show how the water temperature changes over a period of one minute when Jonathan uses this third strategy. Produce a printout of your results and comment on them.

Jonathan decides to install an automatic control mechanism. At 5-second intervals this measures the temperature of the emerging water and adjusts the valve. The adjustment it makes is such that the water temperature at 5-second intervals is given by

$$u_{n+2} = u_{n+1} - \tfrac{1}{2}(u_n - 38).$$

(vii) Build a spreadsheet to show how the water temperature changes over a period of one minute when under this control mechanism, starting at 21°C. Produce a printout of your results and comment on them.

[MEI]

KEY POINTS

1. In iteration, at each stage there is a specific value to use as the starting point for the next calculation. With recursion, you do not have a specific value to work with until you arrive at a final answer. Until then, it is always defined in terms of the next step.

2. The solution to a first-order, linear, homogeneous recurrence relation is of the form:

$$u_n = a^n u_0.$$

3. The solution to a first-order, linear, non-homogeneous recurrence relation with a constant term is of the form:

$$u_n = a^n u_0 + \frac{b(1 - a^n)}{1 - a} \qquad a \neq 1$$

$$u_n = u_0 + nb \qquad\qquad a = 1$$

4. To solve a first-order, linear, non-homogeneous recurrence relation, $u_{n+1} = ku_n + f(n)$, where the final term, $f(n)$, is more complex than a constant, try the step-by-step approach.

$$u_n = ku_{n-1} + f(n) = k^2 u_{n-2} + kf(n-1) + f(n) = \ldots,$$

and then try to sum the series.

5. The solution to a second-order, linear, homogeneous recurrence relation is constructed from the solution(s) to the *auxiliary equation*.

6. The solution to a second-order, linear, non-homogeneous recurrence relation is of the form

$$u_n = \begin{pmatrix} \text{general solution to the associated} \\ \text{homogeneous equation} \end{pmatrix} + \begin{pmatrix} \text{one particular solution} \\ \text{to the full equation} \end{pmatrix}.$$

8   Simulation: spreadsheets and repetitions

As if his whole vocation were endless imitation.

Intimations of immortality, William Wordsworth

The essence of simulation is repetition. This chapter explores why that is and shows one way of achieving repetition without pain. By the end of it you will be able to say whether or not a serving system can cope with a given level of demand, and why it is that a game of tennis can sometimes turn out to be surprisingly one-sided.

❓ In an American game show, the winning contestant is offered a choice of three closed doors, behind one of which is a fabulous prize. The compere knows where the prize is. The contestant chooses a door and the compere then throws open a different door, not the one hiding the prize. The contestant is then offered the option of switching his choice to the other unopened door. Should he switch? Try simulating this with a friend, using three books and a coin.

Simulation modelling

Figure 8.1 illustrates a general simulation model.

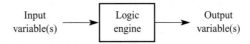

Figure 8.1

The input variables are random variables. For instance, in a simulation of a queuing system they would include the distribution of customer inter-arrival times and the service time distribution.

Clearly you will be dealing here with statistical concepts, but it is sufficient to have an overall feel for the relevant terms, and to know that 'random variable' and 'statistical distribution' have similar meanings. It would be helpful to have worked on simple simulations, transforming, say, two-digit random numbers so as to model discrete distributions.

The logic engine is the mechanism by which the simulation operates on the input variables. It imitates what happens in reality. Thus in a simulation of a queuing system virtual entities (i.e. items such as people or cars) will be created as arrivals using the inter-arrival time distribution; they will be assigned according to some rule to a queue and will wait dutifully for their turn to be served.

Of all mathematical modelling, simulation modelling is the most transparent. A differential equation, modelling a cooling cup of coffee for example, is not at all similar to a cup of coffee. But there is a sense in which a simulation model is more closely identifiable with that which it is modelling.

Finally, there is the output from the simulation. In a queuing simulation you might be interested in, for instance, queuing time, or the length of the queue. These too are random variables. Queuing time varies from entity to entity. Queue length varies over time. For full information you will need to study enough sample observations (entities for queuing time; time intervals for queue length) to be able to infer what the true underlying distributions are. However, distributions can often be summarised by one or two parameters, such as the mean and the standard deviation. Often it is only the mean that is relevant.

It is quite likely, therefore, that your objective will be to build and run a simulation model to estimate a parameter such as the mean queuing time or the mean queue length. To do this you need to know how many repetitions are required.

The statistical knowledge which you need to answer that question is covered in this chapter. But first a spreadsheet approach is developed. This will enable you to repeat a simulation many times.

Simulation mechanisms

In *Decision Mathematics 1*, Chapter 6, simulations were performed by hand, using tables both to summarise the results and to impose the logic. If accuracy requirements demand a very large number of repetitions then this is not a viable approach. Computer implementations will be needed. There are two ways that this can be done, by using a specialist simulation package or by using a spreadsheet.

Specialist computer simulation packages have been available for a long time. In order to be sufficiently flexible the software has had to be relatively unfriendly. Recent advances have led to packages with graphical interfaces. These are certainly easier to use, but tend not to have as much flexibility. However, the fact that tabular layouts help with hand simulations indicates that spreadsheets can be of assistance.

This is demonstrated here by using some of the features and syntax of EXCEL. Other spreadsheet packages have similar features. It is important to be familiar with relative and absolute addressing.

Relative addressing

A relative address in a formula will change when the formula is copied to another cell. If the copying is across columns then the column part of the address will be increased accordingly. If the copying is down rows then the row part will increase accordingly. This is relative addressing.

Absolute addressing

In spreadsheet packages relative addressing is the default, so that an address in a formula will change as described above when the formula is copied across or down. To stop this from happening in EXCEL either the column part of the address, or the row part, or both, is prefixed by $. This fixes that part which follows, and is called absolute addressing.

All spreadsheets use these two forms of addressing. They differ only in their syntax; the logic is the same for all.

There are two formulae in EXCEL which are particularly useful in simulations.

- = LOOKUP(RAND( ), ADDRESS1, ADDRESS2)

    This enables you to simulate a value (a realisation) from a discrete distribution which you can define.

    RAND( ) produces a random number between 0 and 1 or, more formally, a realisation of a uniform random variable defined on the interval 0 to 1.

    ADDRESS1 contains the addresses of the cumulative probabilities of your discrete distribution.

    ADDRESS2 contains the corresponding result, which is delivered into the cell containing the lookup formula.

Figure 8.2 shows an example.

This says where the probability values will be found

This says where the resulting values will be found

	A	B	C
1	=LOOKUP(RAND(),B1:B3,C1:C3)	0	1
2		0.25	2
3		0.75	3

Figure 8.2

The result, which is delivered into A1, depends on the value taken by RAND( ). If the value is between 0 and 0.25 (not including 0.25) then the result 1 is placed in A1. If it is between 0.25 and 0.75 (not including 0.75) the result 2 is returned. If the random number is 0.75 or above the result 3 is returned.

● = IF (CONDITION, VALUE IF TRUE, VALUE IF FALSE)

This logical function allows you to perform most of the logical switching which you need effectively to 'program' the spreadsheet.

CONDITION is a logical condition which may be either true or false.

VALUE IF TRUE is that which is returned to the cell containing the IF formula if the condition is true. VALUE IF FALSE is returned to the cell if the condition is false.

Figure 8.3 shows an example.

	A	B
1	=RAND()	=IF(A1<0.4,1,0)

Figure 8.3

In this example the result delivered into cell B1 depends on the value taken by RAND( ) in cell A1. If that value is between 0 and 0.4 then the result 1 will be returned. If the value is 0.4 or more then the result 0 will be returned.

IF statements can be nested so that the condition can include IF, as can either of the values.

Other logical conditions include:

● = AND (CONDITION1, CONDITION2, ...)

This logical function returns the result TRUE if all of the conditions (there can be as many as you wish) are true. Otherwise it returns the result FALSE.

● = OR (CONDITION1, CONDITION2, ...)

This logical function returns the result TRUE if any (or all) of the conditions are true. If they are all false it returns the result FALSE.

Simulation modelling using a spreadsheet

Simulating a single-server queue system

Single-server queue systems can be dealt with analytically; simulation is not needed. However, any queuing system which is much more complicated does need to be simulated, and building a simulation model of the simplest system is the first step.

The approach is to program a repetition into a single line. If care is taken with absolute addressing that line can then be copied down to provide as many repetitions as are needed. Further repetitions can also be achieved by pressing the recalculation (F9) button.

To set up the simulation the first line usually has to be slightly different from the subsequent, repeated, lines. Figure 8.5 shows the formulae for such a simulation and figure 8.4 shows a simulation of 25 customers. (Note that the inter-arrival time distribution and the service time distribution are held on sheet 2 of the spreadsheet.)

Column A gives customer numbers.

In column B the LOOKUP function is used to simulate the time intervals between successive customers. Thus the results from LOOKUP have to be accumulated to give arrival times.

(In this example the distribution of arrival intervals is that given in the explanation of LOOKUP on page 211.)

In column C the start of service is computed by taking the latest of two times, the time at which the customer arrives and the time at which the server finishes with the previous customer.

In column D LOOKUP is used to simulate the service time for the customer. (In this example, service times are 1, 2 or 3 minutes with probabilities $\frac{1}{6}$, $\frac{1}{2}$ and $\frac{1}{3}$ respectively.)

In column E the time at which the server finishes dealing with the customer is computed.

In columns F and G the formulae compute the time for which the customer queued for service and the mean queuing time for all customers to date.

In columns H and I the formulae compute the time between customers for which the server is idle and the overall server utilisation.

You can see from the output shown in figure 8.4 that, in this situation, the system seems to be coping. Mean queue time is not increasing. It also seems to be making good use of server time, since server utilisation is over 90%.

Run the simulation for more customers to see if the system continues to cope over a longer period of time.

Customer	Arrival time	Start of service	Service time	End of service	Q time	Mean Q length	Server idle	Server utilisation
1	3	3	3	6	0	0.17	0	1.00
2	5	6	2	8	1	0.25	0	1.00
3	7	8	2	10	1	0.20	0	1.00
4	10	10	2	12	0	0.25	0	1.00
5	11	12	3	15	1	0.25	0	1.00
6	13	15	2	17	2	0.33	0	1.00
7	14	17	1	18	3	0.47	0	1.00
8	15	18	2	20	3	0.61	0	1.00
9	18	20	1	21	2	0.65	0	1.00
10	21	21	2	23	0	0.62	0	1.00
11	23	23	2	25	0	0.57	0	1.00
12	25	25	2	27	0	0.52	0	1.00
13	28	28	2	30	0	0.46	1	0.97
14	30	30	3	33	0	0.43	0	0.97
15	32	33	2	35	1	0.42	0	0.97
16	33	35	1	36	2	0.46	0	0.97
17	35	36	2	38	1	0.47	0	0.97
18	37	38	3	41	1	0.47	0	0.98
19	39	41	1	42	2	0.49	0	0.98
20	42	42	1	43	0	0.48	0	0.98
21	45	45	2	47	0	0.44	2	0.94
22	48	48	2	50	0	0.42	1	0.92
23	50	50	2	52	0	0.40	0	0.92
24	53	53	2	55	0	0.38	1	0.91
25	54	55	2	57	1	0.38	0	0.91

Figure 8.4 Output from the spreadsheet shown in figure 8.5

	A	B	C
1	Customer	Arrival time	Start of service
2	1	=LOOKUP(RAND(),Sheet2!A$2:A$4,Sheet2!B$2:B$4)	=B2
3	=A2+1	=B2+LOOKUP(RAND(),Sheet2!A$2:A$4,Sheet2!B$2:B$4)	=MAX(B3,E2)
4	=A3+1	=B3+LOOKUP(RAND(),Sheet2!A$2:A$4,Sheet2!B$2:B$4)	=MAX(B4,E3)

D	E	F
Service time	End of service	Q time
=LOOKUP(RAND(),Sheet2!D$2:D$4,Sheet2!E$2:E$4)	=C2+D2	=C2-B2
=LOOKUP(RAND(),Sheet2!D$2:D$4,Sheet2!E$2:E$4)	=C3+D3	=C3-B3
=LOOKUP(RAND(),Sheet2!D$2:D$4,Sheet2!E$2:E$4)	=C4+D4	=C4-B4

G	H	I
Mean Q length	Server idle	Server utilisation
=SUM(F$2:F3)/C3	=C3-E2	=1-SUM(H$3:H3)/E3
=SUM(F$2:F4)/C4	=C4-E3	=1-SUM(H$3:H4)/E4

Figure 8.5

Simulating a set in tennis

This is a very simple model in which player A wins or loses points to player B according to a fixed probability. The question is how the probability of winning a set is related to the probability of winning a point. Given the complexity of the scoring system, this is not easy to calculate and simulation is the alternative approach. Figure 8.6 on page 217 shows the formulae for a spreadsheet simulation of this and figure 8.7 on page 218 shows a simulation of 50 points.

Column C gives the value of a uniform random variable between 0 and 1.

Column D contains the first player's points score. The general formula (i.e. after the first point) checks to see if a game has just been completed by looking at column J in the previous row. If that cell contains 1 then a game has just been completed and so the random number is ignored and the points score is set to zero. If the cell contains 0 then the random number is checked to see whether this player, or the opponent, has won the point. The points score is increased by 1 if the point is won. (Note that points are recorded as 0, 1, 2, 3, 4, 5, etc., and not as 'love', 15, 30, 40, 'advantage'.)

The formulae in column E perform the same function for the second player as do those in column D for the first player.

Column F contains the first player's game score. The general formula first checks to see if a set has just been completed by looking at column K in the previous row. If that cell contains 1 then a set has just been completed and so the game score is set to zero. If it contains 0 then the point scores are checked to see whether this player has won a game. This happens if the points score is greater than 3 and at least 2 greater than that of the opponent.

The formula in column G does the same for the second player.

Columns H and I contain the set scores. A player's set score is increased if his or her game score is greater than 5 and if he or she has won at least two games more than the other player.

Columns J and K are known as *flags*. They are set to 1 if a game or set, respectively, has just been won, and to 0 otherwise.

Columns J and K could be hidden for the sake of appearances, but they are left on display here for completeness.

Looking at the results, shown in figure 8.7 on page 218, you might think that this game is surprisingly one-sided!

Extend the spreadsheet to simulate more points. Experiment with different values of *p*, so that you can plot the probability of winning a set as a function of the probability of winning a point.

	A	B	C	D	E
1	p=	0.40		Points A	Points B
2				0	0
3			RAND()	=IF(C3<B$1,1+D2,D2)	=IF(C3>B$1,1+E2,E2)
4			RAND()	=IF(J3=1,0),IF(C4<B$1,1+D3,D3))	=IF(J3=1,0),IF(C4>B$1,1+E3,E3)

F	G
Games A	Games B
0	0
=IF(AND(D3>(E3+1),D3>3),1+F2,F2)	=IF(AND(E3>(D3+1),E3>3),1+G2,G2)
=IF(K3=1,0,IF(AND(D4>(E4+1),D4>3),1+F3,F3))	=IF(K3=1,0,IF(AND(E4>(D4+1),E4>3),1+G3,G3))

H	I	J	K
Sets A	Sets B		
0	0		
=IF(AND(F3>(G3+1),F3>5),1+H2,H2)	=IF(AND(G3>(F3+1),G3>5),1+I2,I2)	=IF(OR(F3>F2,G3>G2),1,0)	=IF(OR(H3>H2,I3>I2),1,0)
=IF(AND(F4>(G4+1),F4>5),1+H3,H3)	=IF(AND(G4>(F4+1),G4>5),1+I3,I3)	=IF(OR(F4>F3,G4>G3),1,0)	=IF(OR(H4>H3,I4>I3),1,0)

Figure 8.6

p=	0.40	Points A	Points B	Games A	Games B	Sets A	Sets B		
		0	0	0	0	0	0		
0.53		0	1	0	0	0	0	0	0
0.20		1	1	0	0	0	0	0	0
0.89		1	2	0	0	0	0	0	0
0.61		1	3	0	0	0	0	0	0
0.37		2	3	0	0	0	0	0	0
0.62		2	4	0	1	0	0	1	0
0.39		0	0	0	1	0	0	0	0
0.55		0	1	0	1	0	0	0	0
0.89		0	2	0	1	0	0	0	0
0.76		0	3	0	1	0	0	0	0
0.31		1	3	0	1	0	0	0	0
0.60		1	4	0	2	0	0	1	0
0.23		0	0	0	2	0	0	0	0
0.64		0	1	0	2	0	0	0	0
0.82		0	2	0	2	0	0	0	0
0.46		0	3	0	2	0	0	0	0
0.90		0	4	0	3	0	0	1	0
0.08		0	0	0	3	0	0	0	0
0.55		0	1	0	3	0	0	0	0
0.22		1	1	0	3	0	0	0	0
0.64		1	2	0	3	0	0	0	0
0.41		1	3	0	3	0	0	0	0
0.93		1	4	0	4	0	0	1	0
0.67		0	0	0	4	0	0	0	0
0.67		0	1	0	4	0	0	0	0
0.89		0	2	0	4	0	0	0	0
0.01		1	2	0	4	0	0	0	0
0.72		1	3	0	4	0	0	0	0
0.62		1	4	0	5	0	0	1	0
0.01		0	0	0	5	0	0	0	0
0.00		1	0	0	5	0	0	0	0
0.16		2	0	0	5	0	0	0	0
0.29		3	0	0	5	0	0	0	0
0.37		4	0	1	5	0	0	1	0
0.10		0	0	1	5	0	0	0	0
0.26		1	0	1	5	0	0	0	0
0.88		1	1	1	5	0	0	0	0
0.53		1	2	1	5	0	0	0	0
0.62		1	3	1	5	0	0	0	0
0.32		2	3	1	5	0	0	0	0
0.96		2	4	1	6	0	1	1	1
0.54		0	0	0	0	0	1	0	0
0.30		1	0	0	0	0	1	0	0
0.53		1	1	0	0	0	1	0	0
0.65		1	2	0	0	0	1	0	0
0.31		2	2	0	0	0	1	0	0
0.07		3	2	0	0	0	1	0	0
0.49		3	3	0	0	0	1	0	0
0.08		4	3	0	0	0	1	0	0

Figure 8.7 *Output from the spreadsheet shown in figure 8.6*

1 Generalise the queuing spreadsheet in figure 8.5 to model a single queue, two-server system.

(*Hint*: In this simulation column C should contain 1 if the customer is to be served by the first server, and 2 if by the second server. You can set the value to 1 or 2 by using an IF statement to see which server is available first. Thereafter your computations will depend on whether the value is 1 or 2, so more IF statements will be needed.)

2 Extend the tennis spreadsheet in figure 8.6 to model a men's singles tennis match consisting of best of five sets without tie breaks.

3 Modify the tennis spreadsheet in figure 8.6 to allow for the probability of player A winning a point to depend on whether player A is serving or receiving.

4 Build a spreadsheet model to find a simulated value for π. Do this by generating two uniform random variables between 0 and 1, and computing the sum of their squares. If this is less than 1 then the square root will also be less than 1. This will mean that if you plot a point with co-ordinates corresponding to the two random variables, then that point will be within a quarter circle centred on the origin. If greater than 1 then the point will be within a unit square but outside the quarter circle.

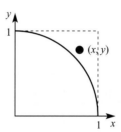

Repeat this many times and compute the ratio of points inside the quarter circle to the total number of points. This will be an estimate of $\frac{\pi}{4}$.

5 A quantity of sand containing 100 pebbles is used to make 100 glass bottles. How many bottles will contain no pebbles? How many will contain 1, 2, 3 or more than 3 pebbles?

Use a spreadsheet to generate a column containing 100 two-digit random variables. (The EXCEL formula "= INT(100*RAND())" will produce a two-digit random variable.)

The first of your two-digit random numbers represents the bottle in which the first pebble ends up. The second two-digit random variable represents the bottle in which the second pebble ends up, and so on.

Count the number of pebbles in bottle number 1. (You can use a spreadsheet column for this.) Count the number of pebbles in bottle number 2, and so on.

Finally, count the number of bottles containing 0, 1, 2, 3 and more than 3 pebbles, again using your spreadsheet.

Repeat your simulation many times and summarise the results.

Simulations sometimes take several repetitions to settle down. This will happen if the initial conditions are not representative in some way. It is therefore common practice to ignore the first few outputs from a simulation to allow for this settling-down period.

Number of repetitions

'How many times should I run my simulation?' Answering this question involves some statistics theory. It also involves using information, like the probabilities in the table below, which is not available to you when you are actually carrying out a simulation. However this all leads up to the procedure on page 222, at the end of this section, which you can use in any practical situation.

Start by thinking of a random variable X which can take certain values. This is going to be the value of the observation you record from your simulation; there will be certain possible values for it to take, each with an associated probability. For example, X might have possible values 1, 2, 3 or 4, with the probabilities shown in the table below.

x	1	2	3	4
$P(X=x)$	$\frac{1}{4}$	$\frac{1}{2}$	$\frac{1}{8}$	$\frac{1}{8}$

❓ Notice that $\frac{1}{4} + \frac{1}{2} + \frac{1}{8} + \frac{1}{8} = 1$. What does this tell you?

The mean, or expected value of X, is given by

$$\mu = E(X) = \sum(x \times P(X=x))$$
$$= \left(1 \times \tfrac{1}{4}\right) + \left(2 \times \tfrac{1}{2}\right) + \left(3 \times \tfrac{1}{8}\right) + \left(4 \times \tfrac{1}{8}\right)$$
$$= \tfrac{17}{8} = 2\tfrac{1}{8}$$

The standard deviation gives you a measure of the spread of the observations of a random variable. It is given by

$$\sigma = \sqrt{\sum[(x-\mu)^2 \times P(X=x)]}.$$

For this distribution this gives

$$\sigma = \sqrt{\left[\left(1-2\tfrac{1}{8}\right)^2 \times \tfrac{1}{4}\right] + \left[\left(2-2\tfrac{1}{8}\right)^2 \times \tfrac{1}{2}\right] + \left[\left(3-2\tfrac{1}{8}\right)^2 \times \tfrac{1}{8}\right] + \left[\left(4-2\tfrac{1}{8}\right)^2 \times \tfrac{1}{8}\right]}$$
$$= 0.927 \text{ (to 3 s.f.)}.$$

So far the meaning of the random variable X has not been defined. From now on it is taken to be the length of a queue at a particular time in a simulation (once it has settled down). It has length 1 with probability $\frac{1}{2}$, length 2 with probability $\frac{1}{4}$, etc. The mean queue length at that time is $2\frac{1}{8}$. (Notice that although the queue length is always a whole number, the mean queue length does not have to be.)

 None of this information is available to you when you are running your simulation and taking regular observations of the length of the queue, perhaps every minute.

Imagine that you make just one simulated observation of the queue length and it is 2. If you were to stop the simulation at that point then you would only have the single observation, 2, from which to infer an estimate of μ. You would then estimate μ to be 2. That would not be too bad, but of course you could have observed a queue length of (say) 4, and that would not have been so good. You can think of the standard deviation as an average error that you might make in using a single observation as an estimate for the mean.

 Usually, like the mean, you do not know the standard deviation and have to estimate it from your observations. The formula for doing this is given as part of step 2 on page 222.

Suppose now that you run two independent repetitions of the simulation, so that you have two independent observations of the queue length (which might be equal, but might not).

 What does the word 'independent' mean?

So you now have a sample of size 2 and you can estimate the mean queue length by adding your two observations together, and dividing by 2.

Since in this theoretical case you know the distribution of X, you can work out the distribution of $Y = \dfrac{X_1 + X_2}{2}$. It is given in the table below.

1	1.5	2	2.5	3	3.5	4
$\frac{1}{16}$	$\frac{2}{8}$	$\frac{1}{4} + \frac{2}{32}$	$\frac{2}{32} + \frac{2}{16}$	$\frac{1}{64} + \frac{2}{16}$	$\frac{2}{64}$	$\frac{1}{64}$
$= \frac{4}{64}$	$= \frac{16}{64}$	$= \frac{20}{64}$	$= \frac{12}{64}$	$= \frac{9}{64}$	$= \frac{2}{64}$	$= \frac{1}{64}$

You can check that this distribution has the same mean of $2\frac{1}{8}$. (Technically, this is saying that the sampling procedure is unbiased.)

However, the calculation of standard deviation does not produce the same result as before:

$$\sqrt{\left[\left(1-2\tfrac{1}{8}\right)^2 \times \tfrac{4}{64}\right] + \left[\left(1.5-2\tfrac{1}{8}\right)^2 \times \tfrac{16}{64}\right] + \ldots + \left[\left(4-2\tfrac{1}{8}\right)^2 \times \tfrac{1}{64}\right]}$$
$$= 0.656 \text{ (to 3 s.f.)}.$$

The standard deviation has been reduced from 0.927 to 0.656.

---

1 What effect has taking a second observation and calculating its mean had upon the probabilities of extreme values?

2 Calling the underlying standard deviation σ, check that the standard deviation for the samples of size 2 is given by $\dfrac{\sigma}{\sqrt{2}}$.

In a similar fashion, if the simulation is repeated n independent times, giving a sample of size n, and the mean and standard deviation are calculated, the mean remains the same but the standard deviation is reduced to $\dfrac{\sigma}{\sqrt{n}}$. This quantity, $\dfrac{\sigma}{\sqrt{n}}$, is known as the *standard error of the mean*.

To complete the statistics background, one more piece of information is required. That is provided by the Central Limit Theorem, which says that the distribution of the mean approaches the Normal distribution ever more closely as n increases. For the Normal distribution, the probability that an observation will be within two standard deviations of the mean is 0.95 (i.e. a 95% chance).

You are now in a position to answer the question of how many repetitions. The procedure runs as follows.

Step 1 Decide on the required accuracy.

Step 2 Run the simulation a few times, 10 or 20 say, to enable you to estimate the value of σ. This estimate is denoted by s. The formula for calculating s from n observations, $x_1, x_2, \ldots, x_n$, is $s = \sqrt{\dfrac{S_{xx}}{n-1}}$ where the sum of squares, S_{xx}, is given by $S_{xx} = \sum\limits_{i=1}^{n} (x_i - \bar{x})^2$ and the sample mean, $\bar{x}$, by $\bar{x} = \dfrac{\sum\limits_{i=1}^{n} x_i}{n}$.

(This may sound rough and ready, but the error in estimating σ decreases very quickly as the sample size increases.)

Step 3 Choose n so that $\dfrac{2s}{\sqrt{n}} <$ the required accuracy (s is the estimate of σ).

(This uses the fact that 95% of observations of a Normally distributed variable lie within two standard deviations of the mean.) The number of repetitions you need is then given by n.

Using the earlier example, estimate the mean queue length to an accuracy of one second.

SOLUTION

You need to choose n such that $2 \times \dfrac{0.927}{\sqrt{n}} \leqslant \dfrac{1}{60}$.

(Notice that the value of $\sigma = 0.927$ was already known in this theoretical example).

This gives $n \geqslant 12\,374.3$. So you would need about $12\,500$ repetitions to achieve that degree of accuracy, which may or may not be feasible, depending on how you are running your simulations.

For an accuracy of half a minute, the computation is $\dfrac{2 \times 0.927}{\sqrt{n}} \leqslant \dfrac{1}{2}$, giving $n \geqslant 13.75$. So for this degree of accuracy only 14 repetitions are needed.

Note

In this case (i.e. looking at queue length) a repetition involves a repeat of the whole simulation up to a particular moment of simulated time. If you were looking at queuing time, then you would have to repeatedly simulate the queuing time of a particular customer. For instance, you might repeatedly simulate the queuing time of the tenth customer, provided that the queue had settled down by that time.

1 A simulation produces the output variable y. The repetitions of the simulation give the following values for y.

221 256 243 240 281 252 270 301 265 265.

Estimate how many repetitions would be needed to produce an estimate of the mean value of y to an accuracy of:

(i) 10

(ii) 5

(iii) 1.

2 A simulation of a queuing system is run to simulate 30 minutes of real time. During that time 20 customers are processed. Their queuing times, which do not include service times, are given in the table.

Customer number	1	2	3	4	5	6	7	8	9	10
Queuing time (minutes)	5	4	0	3	2	5	6	0	0	1

Customer number	11	12	13	14	15	16	17	18	19	20
Queuing time (minutes)	0	4	9	8	7	0	1	2	2	4

(i) Estimate how many repetitions would be needed to produce an estimate of the mean queuing time to an accuracy of:

(a) 1 minute

(b) 0.5 minutes

(c) 0.1 minutes

(ii) Estimate the mean queue length over the 30-minute period by dividing the total time the 20 customers spent queuing by 30.

The standard deviation of the queue length cannot be obtained directly from the simulation output, but is calculated by hand using the arrival and departure times of the each of the 20 customers. It is found to be 1.02.

(iii) Give the accuracy to which the mean queue length will be estimated by performing the computation in part (ii) in each of the three cases given by your answer to part (i).

Verification, validation and variance reduction

Verification is the process of ensuring that a simulation model is doing what is intended. When building a computer program it is easy to make a mistake so a comprehensive error checking procedure is needed. The simulation must be run by hand using the random numbers generated in a computer simulation for a significant number of repetitions.

Validation is the process of ensuring that a simulation model is an adequate model of reality. It checks that what the model was intended to do is appropriate. In most applications of simulation modelling the first step is to build a model of an existing system, so that 'what if' experiments can be conducted on the model, rather than, more expensively, on the real-world system. It is therefore essential to ensure that the model really *is* a good imitation of the existing system. This means checking that the distribution of each output variable is comparable to the corresponding real-world distribution, the means, standard deviations and distribution shapes must be similar.

Variance reduction is the name given to a set of techniques which reduce the variability of the output variable(s), so that higher accuracy can be achieved with fewer repetitions. These techniques are largely statistical in nature and are beyond the scope of this course, but the advantage of using pseudo-random numbers is described below.

ⓔ *Using pseudo-random numbers*

Random numbers generated by a computer are not true random numbers, they are pseudo-random. This means that the same string can be reused. This is particularly useful if you are trying to see whether or not there is a significant difference in the output when some small change is made to the system. Suppose that the output string before the change consists of the numbers $A_1, A_2, A_3, \ldots,$ and that the output string after the change is $B_1, B_2, B_3, \ldots$. The basis for the comparison is to look at the mean of the output before the change and that of the output after the change. If the same pseudo-random numbers have been used in each simulation, then the output strings will tend to be correlated.

This correlation has the effect of reducing the variability of the differences so that the variability of $(A_i - B_i)$ is less than would otherwise have been the case. This is shown in the following formula. (Note that Standard Deviation $= \sqrt{\text{Variance}}$.)

$$\text{Var}(A - B) = \text{Var}(A) - 2 \times \text{Cov}(A, B) + \text{Var}(B)$$

If the variables are positively correlated the covariance is positive

The mean, or expected value, has the property

$$E(A) - E(B) = E(A - B).$$

So to find $E(A) - E(B)$ you can instead find $E(A - B)$.

For instance, suppose that an output variable took the values 12, 15, 9, 17 and 20 over five simulations of a system. Then a small change is made to the system and five runs of the revised simulation using the same pseudo-random numbers produces the results 13, 17, 10, 18 and 20. The improvements are 1, 2, 1, 1 and 0, and these are much less varied than either of the two sets of results.

EXERCISE 8C

1 The scoring system for the game of number tennis involves points, games and sets. When two players are involved the scoring is as follows:

Players compete for points until one player has scored both
- at least two points more than the other, and
- at least four points in all.
The player with the higher points score has then won the game.

Games are contested until one player has won both
- at least two games more than the other, and
- at least six games in all.
The player with the higher games score has then won the set.

When A and B play each other at number tennis, the probability that A wins any particular point is 0.56.

(i) **(a)** Using, row by row, the random numbers at the end of this question, simulate a game between A and B. State your simulation rules carefully. Record full details of the points score throughout the game. State the winner and the final score in the game.
(b) Say how you would estimate the probability of A winning a game.

(ii) (a) Produce a spreadsheet to continue the simulation started in part (i) and complete the simulation of sets of number tennis between A and B. Your spreadsheet should allow the probability of A winning a point to be changed easily.
Give details of the winner of each game and the games score throughout the set.

(b) Which player won the set and what was the final score?

Number tennis is not restricted to two players. When there are three players the rules are amended slightly. The winner of a game must have at least two points more than *each* of the other players, and at least four points in all. The winner of a set must have at least two games more than *each* of the other players, and at least six games in all.

When A, B and C play together at number tennis, the probabilities of each winning any particular point are 0.28, 0.22 and 0.5 respectively.

(iii) (a) Generalise your spreadsheet to simulate games of number tennis between A, B and C. State the new simulation rules and record full details of the game.

(b) What was the final score and which player won the game?

4032	8635	7583	2543	8594	2527
1281	5109	5193	6981	6755	6573
0829	8002	9473	6130	4798	4455
9775	7884	4077	4089	2931	7254
9174	4993	8845	0179	6859	0424
8848	1143	5981	4802	8706	6373
4502	4802	2793	2947	1716	3771
9109	1853	9373	3900	9143	7284

[Oxford, adapted]

2 Twenty companies each have a bay of five reserved spaces in a multi-storey car park (100 spaces in all). Records indicate that at midday, on average, 15% of spaces are empty.

(i) Use a spreadsheet to simulate the presence (1) or absence (0) of a car at midday in each of the 100 spaces. Group your results in sets of five, and thus count how many cars are present for each company. Record this number for each company in a column of your spreadsheet.

(ii) Repeat your simulation ten times and use your results to estimate
(a) the probability that a randomly selected company has all of its spaces occupied
(b) the mean number of spaces occupied per company.

(iii) Calculate the theoretical probability that a randomly selected company has all of its spaces occupied.

[Oxford, adapted]

3 The time taken to serve a customer at a payment till in a department store varies as shown in the following table.

Time to serve (minutes)	1	2	4
Probability	$\frac{1}{4}$	$\frac{1}{2}$	$\frac{1}{4}$

Customers arrive at the till according to the following table.

Time between customers (minutes)	1	2	3	4
Probability	$\frac{1}{6}$	$\frac{1}{3}$	$\frac{1}{3}$	$\frac{1}{6}$

(i)   Build a spreadsheet to simulate the arrival and service of customers at the till.

(ii) Simulate the arrival and serving of 20 customers at the till.

(iii) Find the mean queue length between the time of the arrival of the fifth customer and the start of serving the twentieth customer.
      Why is this more representative of the performance of the system than the mean queue length throughout the simulation?

(iv) Give any times during which the server is not occupied between the time of the arrival of the fifth customer and the time of the arrival of the twentieth customer.
      Find the percentage of the time between the time of the arrival of the fifth customer and the start of serving the twentieth customer, for which the server is occupied.

(v)   Repeat the simulation nine more times. Find the mean of the resulting ten mean queue lengths and estimate its accuracy.

(vi) Use your ten simulations to compute a mean utilisation of the server and to estimate its accuracy.

4 A circular coin of diameter 3 cm is used in a fairground game. Competitors flick the coin along a board marked across with ten parallel lines at 5 cm intervals, and with end sections of width 4 cm. To be valid the coin must land entirely within the scoring region, shown unshaded in the diagram, and it will then count as a win if it does not cross a line. Otherwise it will count as a loss. If it does not land entirely within the scoring area then another attempt is made until it does.

The numbers indicate the distances (in cm) between the lines

x value of centre of coin

Assume that a valid coin is equally likely to land in any position within the scoring region. Thus the *x* value of the centre of a valid coin will be between 1.5 and 51.5.

(i) (a) Give the ranges of *x* values which correspond to the centres of winning coins.

(b) Use a spreadsheet to simulate the *x* values of the centres of ten randomly flicked and valid coins, showing the *x* values of the centres of each of the coins.

(c) Program cells to indicate whether or not each of your simulated coins is in a winning position.

(ii) (a) Repeat your simulation ten times and use your results to estimate the probability of winning.

(b) By considering the regions of the board within which the centre of a winning coin must lie, calculate the probability of winning.

(iii) A new board is to be constructed with ten equally spaced parallel lines 6 cm apart, but with the width of the end sections each being *e* cm. Find the value of *e* for the board to be fair (i.e. one on which the probability of winning is 0.5).

[Oxford, adapted]

5 The diagram shows part of a pinball machine which has tracks along which balls roll. The tracks divide at points at which separators (labelled 1 to 6) deflect the balls one way or the other with equal probabilities.

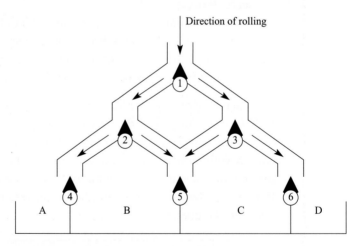

The balls which end up in A or D score five points each. Those which end up in B or C score ten points each.

(i) You are now to use a spreadsheet to simulate the progress of 24 balls. You must indicate clearly:
- how many random numbers you use to simulate the progress of each ball
- the rules which you use to simulate the progress of each ball
- for each ball, a list of the random numbers used and a clear description of the path taken.

(ii) **(a)** What was the mean score per ball from your simulation?

(b) What would you have expected the mean score per ball to be?

[Oxford, adapted]

6 A chocolate factory produces several types of chocolate. 15% have nut centres. The chocolates are mixed together and are randomly packed into cartons of ten.

(i) Two-digit random numbers are to be used to simulate the process of packing chocolates into cartons of ten, distinguishing between nut centres and other chocolates. The rule to be used is that random numbers between 00 and 14 indicate a nut centre.

(a) Use the given rule and the table of two-digit random numbers below to simulate ten cartons of chocolates. Give the number of nut centres in each simulated carton.

Carton										
1	65	16	55	72	04	87	31	29	39	56
2	29	93	95	65	90	95	99	87	46	66
3	36	07	03	49	20	02	59	48	54	35
4	73	34	68	72	44	28	87	44	81	09
5	77	19	52	52	52	65	29	15	82	81
6	23	56	99	82	21	01	62	81	98	14
7	56	32	69	71	27	29	74	87	24	79
8	42	66	10	50	75	40	87	08	26	35
9	84	64	56	47	44	11	22	93	84	75
10	65	26	91	47	67	25	97	25	08	35

(b) Use your results to estimate the probability of a carton containing fewer than two nut centres.

(c) Suppose that the 15% figure quoted is only an approximation and that, in fact, one-seventh of the chocolates have nut centres. Give a simulation rule to simulate the process using two-digit random numbers.

(ii) **(a)** Give a spreadsheet formula which produces the result 1 with probability $\frac{1}{7}$ and the result 0 with probability $\frac{6}{7}$.

(b) Build a spreadsheet to simulate 50 cartons of chocolates when each carton contains ten chocolates, and when one-seventh of the chocolates have nut centres. Your spreadsheet should include 50 rows representing the 50 cartons, each row containing 10 columns to represent the chocolates.

(c) Incorporate in your spreadsheet a count of how many cartons contain fewer than two nut centres and hence estimate the probability of a carton containing fewer than two nut centres.

(iii) A quality control manual states that if the probability of an item from a large batch being faulty is p, then the probability of obtaining fewer than two faulty items in a random sample of size ten is given by:

$$f(p) = (1 + 9p)(1 - p)^9.$$

(a) Apply the formula to produce theoretical answers for the probabilities which you estimated by simulation in parts (i)(b) and (ii)(c).

(b) Comment on any differences between the theoretical results and your simulated results.

(c) Demonstrate how to adapt your spreadsheet to increase the reliability of the simulation results. Give the estimated probability produced by your adaptation.

[MEI, adapted]

7 A newsagent is reviewing her stocking policy. She has records of demand for her newspapers over the past 100 days, and has estimated probabilities from these records. For the Daily Blare they are as shown in the table below.

Daily demand	5	6	7	8	9	10
Probability	0.1	0.2	0.3	0.3	0.05	0.05

The Daily Blare costs 20p per copy from her supplier, and sells for 50p per copy. Unsold copies can be returned to the supplier for a refund of 10p per copy.

(i) Suppose that she orders 6 copies. Show that her profit is £1.80 if the demand is for 8 copies on a particular day. Show that her profit is £1.40 if the demand is for 5 copies.

The newsagent needs to know how many copies of the Daily Blare to order from her supplier each day so as to maximise her profits. A spreadsheet simulation is to be used to find the best number to order.

(ii) Construct a spreadsheet in which column A contains realisations of a uniformly distributed random number between 0 and 1, and column B contains the associated demand for the Daily Blare. Add column C to show the daily profit if the number she orders is 6. Add further columns to show the daily profit if she orders 7, 8, 9, or 10 copies. Produce a printout showing the formulae you used in columns A, B and C. (If you used a 'lookup' formula in column B, then produce a copy of your lookup table as well.)

(iii) Extend your spreadsheet to 100 rows. Produce a printout of the first twenty rows.

(iv) Produce a table showing the estimated mean daily profit for each daily order level.

(v) Estimate the standard error of the mean daily profit when 8 copies are ordered, and hence estimate the accuracy of your simulation.

(vi) Estimate the number of days that would need to be simulated for you to be confident that the mean daily profit when 8 copies are ordered is correct to within 2p.

[MEI]

8 During lunchtimes the time intervals between cars arriving at a roadside restaurant have the distribution given in the table.

Interval (minutes)	1	2	3	4	5	10
Probability	0.5	0.25	0.15	0.05	0.03	0.02

(i) Construct a spreadsheet in which column A contains 100 uniformly distributed random numbers between 0 and 1, and column B contains the associated time intervals between cars arriving at the restaurant. (The first such time interval is to represent the time of arrival of the first car.) Add column C to show the simulated arrival times of cars. Give the formulae you used in columns A, B and C. (If you used a 'lookup' formula in column B, then produce a copy of your lookup table as well.)

The table below gives the distribution of the number of people per car.

Number of people	1	2	3	4
Probability	0.4	0.3	0.2	0.1

(ii) Add column(s) to your spreadsheet to simulate the number of people arriving in each car. Give the formulae which you use.

(iii) Use the 'IF' function to produce a final column in which an entry contains the number of people in the corresponding car if the arrival time of the car is 120 minutes or less, and 0 otherwise. Place a formula giving the sum of the numbers in this column into a convenient cell. Give the formulae which you use.

(iv) Your spreadsheet simulates the number of people arriving at the restaurant in the two-hour lunchtime period. Run your simulation 10 times and produce a table of results. Compute the mean number of people per simulation run.

(v) From your results calculate an estimate of the standard error of the mean number of people arriving over the two-hour period. Estimate the number of simulation runs that are required for you to be confident that the mean number of people given by the simulation is correct to within 5 people.

[MEI]

9 (i) Into the first row of the first column of a spreadsheet enter a formula to give a uniformly distributed random number between 0 and 1. Repeat for the first row of the second column. In the first row of the third column of your spreadsheet enter a formula to give the sum of the squares of your two random numbers. In the first row of the fourth column enter a formula to give the result 1 if the sum of the squares is less than 1, and 0 otherwise. Print out your formulae.

(ii) Copy down your four columns for 1000 rows. Create a cell containing 0.004 times the sum of the entries in the fourth column. Print out the formula for this cell and its value.

(iii) The value which you obtained in part (ii) should be a simulated estimate of π. By regarding your two random numbers as the x and y co-ordinates of a point in the plane, explain why this is.

(iv) Repeat your simulation 12 times by using the recalculation facility of your spreadsheet, recording the 12 simulated values of π. Find the mean and standard deviation of your simulated values.

(v) Use the standard deviation which you computed in part (iv) to compute an estimate of the number of times which you need to repeat the simulation so that you can be confident that the mean value of your simulated values of π is correct to within 0.0005.

(vi) An alternative method to simulate a value of π uses three uniformly distributed random variables between 0 and 1. The sum of the squares of these is found, and the result is compared to 1. This is repeated 1000 times, and the number of results less than 1 is multiplied by 0.006.

Build a spreadsheet to simulate an estimate of π using this method. Use it to produce 12 estimates of π. Investigate whether or not the method seems to improve on the earlier method which used only two random variables.

[MEI]

10 Jorge wishes to cross a single-carriageway road, with one lane in each direction, at the point X, as shown below.

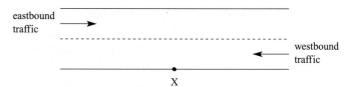

Westbound vehicles pass X with a between-vehicle time interval given by the distribution in the table.

Between-vehicle interval (seconds)	3	6	9	12
Probability	0.4	0.3	0.2	0.1

Eastbound vehicles have the same distribution for between-vehicle time intervals.

Throughout this question, the time taken for Jorge to cross the road is his waiting time plus his crossing time.

(i) Jorge takes 8 seconds to walk across the road. He needs to wait until there is an 8-second interval with no cars passing X before he can begin to cross. He arrives at X as a vehicle passes X in each direction.

A simulation of the times after his arrival at which vehicles pass X produces the results shown in the table below.

Time at which westbound vehicles pass X (seconds)	3	9	12	18	27	30	42	45	51	60
Time at which eastbound vehicles pass X (seconds)	9	15	24	27	30	39	42	45	57	69

(a) Show that it takes Jorge 38 seconds to cross the road in the above simulation.

(b) Create an EXCEL spreadsheet to simulate vehicle passing times in both directions. Give the formulae which you used and print out your times. Find how long it takes for Jorge to cross the road in your simulated situation.

(c) Repeat your simulation a further seven times. Produce a printout showing your simulations, and give a table showing all nine times for Jorge to cross the road.

(d) Compute an estimate of how long on average it will take for Jorge to cross the road.

(e) Compute an estimate of how many times the simulation would have to be run for you to be confident that the estimate of Jorge's average time to cross the road is accurate to within 1 second.

(ii) The County Council decide to make it easier for people to cross the road at X by installing a traffic island.

eastbound traffic

westbound traffic

X

Jorge now has to cross one lane at a time. It takes him 4 seconds to cross each lane. It takes Jorge no time to cross the island, and when he arrives at X a vehicle has just passed in each direction.

(a) Using the simulated times given in part (i), show that it now takes Jorge 13 seconds to cross the road.

(b) Using your simulated times from part (i)(b), find out how long it now takes Jorge to cross the road.

(c) Using your simulated times from part (i)(c), complete another seven simulations for the time needed for Jorge to cross the road using the traffic island. Tabulate all nine results.

(d) Compute an estimate of how long on average it will take for Jorge to cross the road using the traffic island.

(e) Compute an estimate of the accuracy of your estimate of Jorge's average time to cross the road when using the traffic island.

(iii) State any conclusions which you can be sure of as a result of your simulations.

[MEI]

11 In a closed community, x individuals have an infection at time t hours and y individuals are free from infection. The number of individuals in the community, n, is constant, so that $x + y = n$ at all times t. When an individual catches the infection, x increases by 1 and y decreases by 1. There is no recovery from the infection.

Each hour, either one person catches the infection or nobody does. The probability of one person catching the infection in the hour beginning at time t is $0.02xy$.

(i) Given that at time $t = 0$, $x = 1$ and $y = 5$, build a spreadsheet to simulate the progression of the infection through the community. At hourly intervals compare a random number between 0 and 1 to the current value of $0.02xy$ to decide whether or not an infection occurs in that interval. Run your simulation to find out how long it takes before all members of the community are infected. Print out your results and give the time taken.

(ii) Repeat your simulation ten times. List your ten results for the times taken before all of the community is infected. Use these data to produce an estimate of how many simulations would be needed to estimate the expected time to within an accuracy of 10% with 95% confidence. Show the formulae you use in your calculations.

In a similar community, there is an additional possibility of recovery from the infection. Once recovered, an individual cannot be re-infected. The number of individuals at time t who have been infected and have recovered is z, so that $x + y + z = n$ at all times t.

In the hour beginning at time t either one infected person recovers, with probability $0.1x$, or no-one recovers.

(iii) Adapt your simulation to model the progress of the infection. Run your simulation and include a printout.

(iv) Run your simulation ten times and produce a summary of your results. Describe what happens.

[MEI]

1 A simulation model consists of input random variable(s), a logic engine and output random variable(s).

2 Spreadsheet formulae can include *relative addresses* and *absolute addresses*. In EXCEL, $ makes absolute that part of an address immediately following it.

3 Spreadsheet functions in EXCEL which are useful in simulation modelling include LOOKUP, RAND(), IF, AND and OR.

4 The number of repetitions needed is determined by considering the standard error of the mean, $\frac{\sigma}{\sqrt{n}}$. A rule of thumb is to choose n so that $\frac{2s}{\sqrt{n}} <$ the required accuracy, where s is an estimate of σ.

5 *Verification* is the process of ensuring that a simulation model is doing what the designer intended.

6 *Validation* is the process of checking a simulation model is an adequate model of reality, i.e. that what the designer intended was correct.

7 *Variance reduction* techniques reduce the variability of the output variable(s).

Answers

Chapter 1

❓ (Page 1)

1 There are ten integer (lattice) points in the feasible region.

2 $(2, 6)$ with a value of 22.

3 $(1, 2, 17)$ with value 56.

4 The problem is three-dimensional and it is difficult to draw a three-dimensional graph.

5 You might have solved this by looking at all of the feasible points or you might have made z as large as possible given that y has to be at least 2 and x at least 1. It's not always as easy as that!

❓ (Page 4)

See text that follows.

❓ (Page 7)

A zero entry in the pivot column means that the pivot variable does not appear in the constraint represented by that row. Thus there is no limit on how much the variable can be increased as far as that constraint is concerned. A negative entry means that the coefficient of the pivot element is negative. Thus increases in the value of the pivot element have to be balanced by *increasing* the value of the row's basic element. There is no limit on such increases.

❓ (Page 10)

See text that follows.

❓ (Page 11)

See text that follows.

Exercise 1A (Page 11)

1

I	x	y	s_1	s_2	s_3	RHS
1	−1	−0.8	0	0	0	0
0	1	1	1	0	0	1000
0	2	1	0	1	0	1500
0	3	2	0	0	1	2400
1	−0.2	0	0.8	0	0	800
0	1	1	1	0	0	1000
0	1	0	−1	1	0	500
0	1	0	−2	0	1	400
1	0	0	0.4	0	0.2	880
0	0	1	3	0	−1	600
0	0	0	1	1	−1	100
0	1	0	−2	0	1	400

Solution: $x = 400$, $y = 600$, $I = 880$.

2	P	x	y	s_1	s_2	s_3	RHS
	1	−16	−24	0	0	0	0
	0	2	3	1	0	0	24
	0	2	1	0	1	0	16
	0	0	1	0	0	1	6
	1	−16	0	0	0	24	144
	0	2	0	1	0	−3	6
	0	2	0	0	1	−1	10
	0	0	1	0	0	1	6
	1	0	0	8	0	0	192
	0	1	0	0.5	0	−1.5	3
	0	0	0	−1	1	2	4
	0	0	1	0	0	1	6

Solution: $x = 3$, $y = 6$, $P = 192$.

Note that there is an alternative solution in this case, which emerges by choosing the x column for the first pivot. This is because the objective function is parallel to one of the constraint lines.

P	x	y	s_1	s_2	s_3	RHS
1	−16	−24	0	0	0	0
0	2	3	1	0	0	24
0	2	1	0	1	0	16
0	0	1	0	0	1	6
1	0	−16	0	8	0	128
0	0	2	1	−1	0	8
0	1	0.5	0	0.5	0	8
0	0	1	0	0	1	6
1	0	0	8	0	0	192
0	0	1	0.5	−0.5	0	4
0	1	0	−0.25	0.75	0	6
0	0	0	−0.5	0.5	1	2

Solution: $x = 6$, $y = 4$, $P = 192$.

3	P	x	y	z	s_1	s_2	RHS
	1	−9	−10	−6	0	0	0
	0	2	3	4	1	0	3
	0	6	6	2	0	1	8
	1	$-\frac{7}{3}$	0	$\frac{22}{3}$	$\frac{10}{3}$	0	10
	0	$\frac{2}{3}$	1	$\frac{4}{3}$	$\frac{1}{3}$	0	1
	0	2	0	−6	−2	1	2
	1	0	0	$\frac{1}{3}$	1	$\frac{7}{6}$	$\frac{37}{3}$
	0	0	1	$\frac{10}{3}$	1	$-\frac{1}{3}$	$\frac{1}{3}$
	0	1	0	−3	−1	0.5	1

Solution: $x = 1$, $y = \frac{1}{3}$, $z = 0$, $P = 12\frac{1}{3}$.

4

P	w	x	y	z	s_1	s_2	RHS
1	−3	−2	0	0	0	0	0
0	1	1	1	1	1	0	150
0	2	1	3	4	0	1	200
1	0	−0.5	4.5	6	0	1.5	300
0	0	0.5	−0.5	−1	1	−0.5	50
0	1	0.5	1.5	2	0	0.5	100
1	0	0	4	5	1	1	350
0	0	1	−1	−2	2	−1	100
0	1	0	2	3	−1	1	50

Solution: $w = 50$, $x = 100$, $y = 0$, $z = 0$, $P = 350$.

5

P	w	x	y	z	s_1	s_2	s_3	RHS
1	−3	−2	0	0	0	0	0	0
0	1	1	1	1	1	0	0	150
0	2	1	3	4	0	1	0	200
0	−1	1	0	0	0	0	1	0
1	0	−0.5	4.5	6	0	1.5	0	300
0	0	0.5	−0.5	−1	1	−0.5	0	50
0	1	0.5	1.5	2	0	0.5	0	100
0	0	1.5	1.5	2	0	0.5	1	100
1	0	0	5	$\frac{20}{3}$	0	$\frac{5}{3}$	$\frac{1}{3}$	$\frac{1000}{3}$
0	0	0	−1	$-\frac{5}{3}$	1	$-\frac{2}{3}$	$-\frac{1}{3}$	$\frac{50}{3}$
0	1	0	1	$\frac{4}{3}$	0	$\frac{1}{3}$	$-\frac{1}{3}$	$\frac{200}{3}$
0	0	1	1	$\frac{4}{3}$	0	$\frac{1}{3}$	$\frac{2}{3}$	$\frac{200}{3}$

Solution: $w = 66\frac{2}{3}$, $x = 66\frac{2}{3}$, $y = 0$, $z = 0$, $P = 333\frac{1}{3}$.

Activity (Page 15)

P	x	y	s_1	s_2	s_3	RHS
1	−8	−6	0	0	0	0
0	4	1	1	0	0	40
0	4	3	0	1	0	48
0	20	31	0	0	1	400
1	0	−4	2	0	0	80
0	1	0.25	0.25	0	0	10
0	0	2	−1	1	0	8
0	0	16	0	−5	1	160
1	0	0	0	2	0	96
0	1	0	$\frac{3}{8}$	$-\frac{1}{8}$	0	9
0	0	1	−0.5	0.5	0	4
0	0	0	8	−13	1	96

Solution: $x = 9$, $y = 4$, $P = 96$.

Pivoting again (on the s_1 column):

1	0	0	0	2	0	96
0	1	0	0	$\frac{31}{64}$	$-\frac{3}{64}$	4.5
0	0	1	0	$-\frac{5}{16}$	$\frac{1}{16}$	10
0	0	0	1	$-\frac{13}{8}$	$\frac{1}{8}$	12

Solution: $x = 4.5$, $y = 10$, $P = 96$.

Exercise 1B (Page 23)

1

Q	I	x	y	s_1	s_2	s_3	s_4	a	RHS
1	0	1	1	0	0	0	−1	0	800
0	1	−1	−0.8	0	0	0	0	0	0
0	0	1	1	1	0	0	0	0	1000
0	0	2	1	0	1	0	0	0	1500
0	0	3	2	0	0	1	0	0	2400
0	0	1	1	0	0	0	−1	1	80

Q	I	x	y	s_1	s_2	s_3	s_4	a	RHS
1	0	0	0	0	0	0	0	−1	0
0	1	−0.2	0	0	0	0	−0.8	0.8	640
0	0	0	0	1	0	0	1	−1	200
0	0	1	0	0	1	0	1	−1	700
0	0	1	0	0	0	1	2	−2	800
0	0	1	1	0	0	0	−1	1	800

Q	I	x	y	s_1	s_2	s_3	s_4	a	RHS
	1	−0.2	0	0.8	0	0	0		800
	0	0	0	1	0	0	1		200
	0	1	0	−1	1	0	0		500
	0	1	0	−2	0	1	0		400
	0	1	1	1	0	0	0		1000

Q	I	x	y	s_1	s_2	s_3	s_4	a	RHS
	1	0	0	0.4	0	0.2	0		880
	0	0	0	1	0	0	1		200
	0	0	0	1	1	−1	0		100
	0	1	0	−2	0	1	0		400
	0	0	1	3	0	−1	0		600

Solution: $x = 400$, $y = 600$, $I = 880$.

2

Q	P	x	y	s_1	s_2	s_3	s_4	a	RHS
1	0	0	1	0	0	0	−1	0	2
0	1	−16	−24	0	0	0	0	0	0
0	0	2	3	1	0	0	0	0	24
0	0	2	1	0	1	0	0	0	16
0	0	0	1	0	0	1	0	0	6
0	0	0	1	0	0	0	−1	1	2
1	0	0	0	0	0	0	0	−1	0
0	1	−16	0	0	0	0	−24	24	48
0	0	2	0	1	0	0	3	−3	18
0	0	2	0	0	1	0	1	−1	14
0	0	0	0	0	0	1	1	−1	4
0	0	0	1	0	0	0	−1	1	2
	1	−16	0	0	0	24	0		144
	0	2	0	1	0	−3	0		6
	0	2	0	0	1	−1	0		10
	0	0	0	0	0	1	1		4
	0	0	1	0	0	1	0		6
	1	0	0	8	0	0	0		192
	0	1	0	0.5	0	−1.5	0		3
	0	0	0	−1	1	2	0		4
	0	0	0	0	0	1	1		4
	0	0	1	0	0	1	0		6

Solution: $x = 3$, $y = 6$, $P = 192$.

3

Q	P	x	y	z	s_1	s_2	s_3	a	RHS
1	0	1	1	1	0	0	−1	0	1
0	1	−9	−10	−6	0	0	0	0	0
0	0	2	3	4	1	0	0	0	3
0	0	6	6	2	0	1	0	0	8
0	0	1	1	1	0	0	−1	1	1
1	0	0	0	0	0	0	0	−1	0
0	1	0	−1	3	0	0	−9	9	9
0	0	0	1	2	1	0	2	−2	1
0	0	0	0	−4	0	1	6	−6	2
0	0	1	1	1	0	0	−1	1	1
	1	0	0	5	1	0	−7		10
	0	0	1	2	1	0	2		1
	0	0	0	−4	0	1	6		2
	0	1	0	−1	−1	0	−3		0
	1	0	0	$\frac{1}{3}$	1	$\frac{7}{6}$	0		$\frac{37}{3}$
	0	0	1	$\frac{10}{3}$	1	$-\frac{1}{3}$	0		$\frac{1}{3}$
	0	0	0	$-\frac{2}{3}$	0	$\frac{1}{6}$	1		$\frac{1}{3}$
	0	1	0	−3	−1	0.5	0		1

Solution: $x = 1$, $y = \frac{1}{3}$, $P = 12\frac{1}{3}$.

4

Q	P	w	x	y	z	s_1	s_2	s_3	a	RHS
1	0	0	0	1	1	0	0	−1	0	50
0	1	−3	−2	0	0	0	0	0	0	0
0	0	1	1	1	1	1	0	0	0	150
0	0	2	1	3	4	0	1	0	0	200
0	0	0	0	1	1	0	0	−1	1	50
1	0	0	0	0	0	0	0	0	−1	0
0	1	−3	−2	0	0	0	0	0	0	0
0	0	1	1	0	0	1	0	1	−1	100
0	0	2	1	0	1	0	1	3	−3	50
0	0	0	0	1	1	0	0	−1	1	50
	1	0	−0.5	0	1.5	0	1.5	4.5		75
	0	0	0.5	0	−0.5	1	−0.5	−0.5		75
	0	1	0.5	0	0.5	0	0.5	1.5		25
	0	0	0	1	1	0	0	−1		50
	1	1	0	0	2	0	2	6		100
	0	−1	0	0	−1	1	−1	−2		50
	0	2	1	0	1	0	1	3		50
	0	0	0	1	1	0	0	−1		50

Solution: $w = 0$, $x = 50$, $y = 50$, $z = 0$, $P = 100$.

5

Q	P	w	x	y	z	s_1	s_2	s_3	s_4	a	RHS
1	0	0	0	1	1	0	0	−1	0	0	50
0	1	−3	−2	0	0	0	0	0	0	0	0
0	0	1	1	1	1	1	0	0	0	0	150
0	0	2	1	3	4	0	1	0	0	0	200
0	0	−1	1	0	0	0	0	0	1	0	0
0	0	0	0	1	1	0	0	−1	0	1	50
1	0	0	0	0	0	0	0	0	0	−1	0
0	1	−3	−2	0	0	0	0	0	0	0	0
0	0	1	1	0	0	1	0	1	0	−1	100
0	0	2	1	0	1	0	1	3	0	−3	50
0	0	−1	1	0	0	0	0	0	1	0	0
0	0	0	0	1	1	0	0	−1	0	1	50
	1	0	−0.5	0	1.5	0	1.5	4.5	0		75
	0	0	0.5	0	−0.5	1	−0.5	−0.5	0		75
	0	1	0.5	0	0.5	0	0.5	1.5	0		25
	0	0	1.5	0	0.5	0	0.5	1.5	1		25
	0	0	0	1	1	0	0	−1	0		50
	1	0	0	0	$\frac{5}{3}$	0	$\frac{5}{3}$	5	$\frac{1}{3}$		$\frac{250}{3}$
	0	0	0	0	$-\frac{2}{3}$	1	$-\frac{2}{3}$	−1	$-\frac{1}{3}$		$\frac{200}{3}$
	0	1	0	0	$\frac{1}{3}$	0	$\frac{1}{3}$	1	$-\frac{1}{3}$		$\frac{50}{3}$
	0	0	1	0	$\frac{1}{3}$	0	$\frac{1}{3}$	1	$\frac{2}{3}$		$\frac{50}{3}$
	0	0	0	1	1	0	0	−1	0		50

Solution: $w = 16\frac{2}{3}$, $x = 16\frac{2}{3}$, $y = 50$, $z = 0$, $P = 83\frac{1}{3}$.

Activity (Page 23)

Q	P	x	y	s_1	s_2	s_3	a_1	a_2	RHS
1	0	3	8	0	−1	−1	0	0	461
0	1	−3	−2	0	0	0	0	0	0
0	0	2	5	1	0	0	0	0	300
0	0	3	7	0	−1	0	1	0	441
0	0	0	1	0	0	−1	0	1	20
1	0	3	0	0	−1	7	0	−8	301
0	1	−3	0	0	0	−2	0	2	40
0	0	2	0	1	0	5	0	−5	200
0	0	3	0	0	−1	7	1	−7	301
0	0	0	1	0	0	−1	0	1	20
1	0	0.2	0	−1.4	−1	0	0	−1	21
0	1	−2.2	0	0.4	0	0	0	0	120
0	0	0.4	0	0.2	0	1	0	−1	40
0	0	0.2	0	−1.4	−1	0	1	0	21
0	0	0.4	1	0.2	0	0	0	0	60
1	0	0	0	−1.5	−1	−0.5	0	−0.5	1
0	1	0	0	1.5	0	5.5	0	−5.5	340
0	0	1	0	0.5	0	2.5	0	−2.5	100
0	0	0	0	−1.5	−1	−0.5	1	0.5	1
0	0	0	1	0	0	−1	0	1	20

No further reduction in Q is possible, therefore there is no feasible solution to the problem.

Exercise 1C (Page 26)

1

I	x	y	s_1	s_2	s_3	s_4	a	RHS
1	−(1 + M)	−(0.8 + M)	0	0	0	M	0	−800M
0	1	1	1	0	0	0	0	1000
0	2	1	0	1	0	0	0	1500
0	3	2	0	0	1	0	0	2400
0	1	1	0	0	0	−1	1	800
1	−0.2	0	0	0	0	−0.8	0.8 + M	640
0	0	0	1	0	0	1	−1	200
0	1	0	0	1	0	1	−1	700
0	1	0	0	0	1	2	−2	800
0	1	1	0	0	0	−1	1	800
1	−0.2	0	0.8	0	0	0	M	800
0	0	0	1	0	0	1	−1	200
0	1	0	−1	1	0	0	0	500
0	1	0	−2	0	1	0	0	400
0	1	1	1	0	0	0	0	1000
1	0	0	0.4	0	0.2	0	M	880
0	0	0	1	0	0	1	−1	200
0	0	0	1	1	−1	0	0	100
0	1	0	−2	0	1	0	0	400
0	0	1	3	0	−1	0	0	600

Solution: $x = 400$, $y = 600$, $I = 880$.

2

P	x	y	s_1	s_2	s_3	s_4	a	RHS
1	−16	−(24 + M)	0	0	0	M	0	−2M
0	2	3	1	0	0	0	0	24
0	2	1	0	1	0	0	0	16
0	0	1	0	0	1	0	0	6
0	0	1	0	0	0	−1	1	2
1	−16	0	0	0	0	−24	24 + M	48
0	2	0	1	0	0	3	−3	18
0	2	0	0	1	0	1	−1	14
0	0	0	0	0	1	1	−1	4
0	0	1	0	0	0	−1	1	2
1	−16	0	0	0	24	0	M	144
0	2	0	1	0	−3	0	0	6
0	2	0	0	1	−1	0	0	10
0	0	0	0	0	1	1	−1	4
0	0	1	0	0	1	0	0	6
1	0	0	8	0	0	0	M	192
0	1	0	0.5	0	−1.5	0	0	3
0	0	0	−1	1	2	0	0	4
0	0	0	0	0	1	1	−1	4
0	0	1	0	0	1	0	0	6

Solution: $x = 3$, $y = 6$, $P = 192$.

3

P	x	y	z	s_1	s_2	s_3	a	RHS
1	−(9 + M)	−(10 + M)	−(6 + M)	0	0	M	0	−M
0	2	3	4	1	0	0	0	3
0	6	6	2	0	1	0	0	8
0	1	1	1	0	0	−1	1	1
1	1	0	4	0	0	−10	10+M	10
0	−1	0	1	1	0	3	−3	0
0	0	0	−4	0	1	6	−6	2
0	1	1	1	0	0	−1	1	1
1	$-\frac{7}{3}$	0	$\frac{22}{3}$	$\frac{10}{3}$	0	0	M	10
0	$-\frac{1}{3}$	0	$\frac{1}{3}$	$\frac{1}{3}$	0	1	−1	0
0	2	0	−6	−2	1	0	0	2
0	$\frac{2}{3}$	1	$\frac{4}{3}$	$\frac{1}{3}$	0	0	0	1
1	0	0	$\frac{1}{3}$	1	$\frac{7}{6}$	0	M	$\frac{37}{3}$
0	0	0	$-\frac{2}{3}$	0	$\frac{1}{6}$	1	−1	$\frac{1}{3}$
0	1	0	−3	−1	0.5	0	0	1
0	0	1	$\frac{10}{3}$	1	$-\frac{1}{3}$	0	0	$\frac{1}{3}$

Solution: $x = 1$, $y = \frac{1}{3}$, $P = 12\frac{1}{3}$.

4

P	w	x	y	z	s_1	s_2	s_3	a	RHS
1	-3	-2	-M	-M	0	0	M	0	-50M
0	1	1	1	1	1	0	0	0	150
0	2	1	3	4	0	1	0	0	200
0	0	0	1	1	0	0	-1	1	50
1	-3	-2	0	0	0	0	0	M	0
0	1	1	0	0	1	0	1	-1	100
0	2	1	0	1	0	1	3	-3	50
0	0	0	1	1	0	0	-1	1	50
1	0	-0.5	0	1.5	0	1.5	4.5	M − 4.5	75
0	0	0.5	0	-0.5	1	-0.5	-0.5	0.5	75
0	1	0.5	0	0.5	0	0.5	1.5	-1.5	25
0	0	0	1	1	0	0	-1	1	50
1	1	0	0	2	0	2	6	M − 6	100
0	-1	0	0	-1	1	-1	-2	2	50
0	2	1	0	1	0	1	3	-3	50
0	0	0	1	1	0	0	-1	1	50

Solution: $w = 0$, $x = 50$, $y = 50$, $z = 0$, $P = 100$.

5

P	w	x	y	z	s_1	s_2	s_3	s_4	a	RHS
1	-3	-2	-M	-M	0	0	M	0	0	-50M
0	1	1	1	1	1	0	0	0	0	150
0	2	1	3	4	0	1	0	0	0	200
0	-1	1	0	0	0	0	0	1	0	0
0	0	0	1	1	0	0	-1	0	1	50
1	-3	-2	0	0	0	0	0	0	M	0
0	1	1	0	0	1	0	1	0	-1	100
0	2	1	0	1	0	1	3	0	-3	50
0	-1	1	0	0	0	0	0	1	0	0
0	0	0	1	1	0	0	-1	0	1	50
1	0	-0.5	0	1.5	0	1.5	4.5	0	M − 4.5	75
0	0	0.5	0	-0.5	1	-0.5	-0.5	0	0.5	75
0	1	0.5	0	0.5	0	0.5	1.5	0	-1.5	25
0	0	1.5	0	0.5	0	0.5	1.5	1	-1.5	25
0	0	0	1	1	0	0	-1	0	1	50
1	0	0	0	$\frac{5}{3}$	0	$\frac{5}{3}$	5	$\frac{1}{3}$	M − 5	$\frac{250}{3}$
0	0	0	0	$-\frac{2}{3}$	1	$-\frac{2}{3}$	-1	$-\frac{1}{3}$	1	$\frac{200}{3}$
0	1	0	0	$\frac{1}{3}$	0	$\frac{1}{3}$	1	$-\frac{1}{3}$	-1	$\frac{50}{3}$
0	0	1	0	$\frac{1}{3}$	0	$\frac{1}{3}$	1	$\frac{2}{3}$	-1	$\frac{50}{3}$
0	0	0	1	1	0	0	-1	0	1	50

Solution: $w = 16\frac{2}{3}$, $x = 16\frac{2}{3}$, $y = 50$, $z = 0$, $P = 83\frac{1}{3}$.

Exercise 1D (Page 32)

1 $x = 6$, $y = 14$, value $= 26$

2 (i) $x = 3$, $y = 0$, $z = 0$, value $= 9$
   (ii) $x = 2$, $y = 0$, $z = 1$, value $= 7$

3 (i) $x = 12$, $y = 0$, value $= 36$
   (ii) infeasible
   (iii) The minimum value for the sum of the artificial variables is 4.

4 (i) $x = 7$, $y = 40$, value $= 54$
   (ii) $x = 10$, $y = 10$, value $= 30$
   (iii) $x = 10$, $y = 10$, value $= 30$

5 unbounded

6 Let x tonnes of X be produced and y tonnes of Y.
   (i) $200x + 100y \leqslant 1000$ (finance constraint)
      $8x + 8y \leqslant 48$ (staff constraint)
      $x + 3y \leqslant 15$ (space constraint)
      $x \geqslant 0$ and $y \geqslant 0$
   (ii) $P = 160x + 120y$
   (iii), (iv) $x = 4$, $y = 2$, $P = 880$

7 Let m be the number of hours on the moped.
   Let c be the number of hours in the car.
   Let l be the number of hours in the lorry.
   (i) Maximise $20m + 40c + 30l$
      Subject to $m + c + l \leqslant 3$
      $$\frac{m}{3} + c + \frac{3l}{2} \leqslant 2.5$$
      $20m \leqslant 55$
      $40c + 30l \leqslant 55$
   (ii), (iii) Two iterations give $m = 1.625$, $c = 1.375$, $l = 0$, covering 87.5 miles.
   (iv) Yes.
      Put the moped in the car and drive for 1.375 hours, covering 55 miles. Then use the moped for the remaining 1.625 hours covering a further 32.5 miles.
      All of the time is used and $1\frac{11}{12}$ gallons of petrol are used.

8 (i)

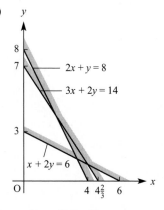

   (ii) There is no evident feasible solution (i.e. no feasible basis).
   (iii) (a) $x = 0$, $y = 8$ and $C = 8$.
      $2x + y = 8$ which is $\geqslant 8$.
      $3x + 2y = 16$ which is $\geqslant 14$.
      $x + 2y = 16$ which is $\geqslant 6$.
   (b) $s_1 = 0$. $2x + y = 8$, so there is no slack in this constraint.
      $s_2 = 2$. $3x + 2y = 16$, which is 2 in excess of the minimum.
      $s_3 = 10$. $x + 2y = 16$, which is 10 in excess of the minimum.
   (iv) Optimum is $C = 5$, given by $x = 4$ and $y = 1$.

9 (i) $x = 3$, $y = 5$, $P = 38$
   (ii) (a) $(4, 3)$
   (b) $P = 6x + 4y = 6(4 + s_3 - s_4) + 4(3 - 3s_3 + 2s_4) = 36 - 6s_3 + 2s_4$
      P can be increased further by increasing s_4.
   (iii) (a) $s_1 = 1$, $s_2 = 0$, $s_3 = 0$, $s_4 = 1$

(b) Inequality 1 has a slack of 1, i.e. $x + 2y = 13$, one less than the maximum.
Inequalities 2 and 3 are tight.
Inequality 4 has a slack of 1, i.e. $3x + y = 14$, one less than the maximum.

10 (i) Return $= 0.11a + 0.15b + 0.08c$
$\qquad = 0.11a + 0.15b + 0.08(1 - a - b)$
$\qquad = 0.08 + 0.03a + 0.07b$
$\quad$ Risk $= 0.9 + 0.1a + 0.3b$

(ii) $a \leqslant \frac{1}{2}, b \leqslant \frac{1}{4}, c \geqslant \frac{1}{3}$

(iii) $a = \frac{1}{2}, b = \frac{1}{14}, c = \frac{3}{7}$

(iv) Finding an initial feasible solution.

11 (i) $x = 17, y = 7, P = 31$

(ii) Iteration 1: solution $= (0, 9)$
$\quad$ Iteration 2: solution $= (5\frac{2}{3}, 9)$

(iii) $s_1 = 0, s_2 = 0, s_3 = 2, s_4 = 3$

12 (i) Let t be the number of tables produced.
$\quad$ Let c be the number of chairs produced.
$\quad$ Maximise $80c + 350t$
$\quad$ Subject to $20c + 100t \leqslant 10\,000$
$\qquad\qquad\quad 4c + 15t \ \ \leqslant 1950$

(ii) $c = 450, t = 10$, income $= £39\,500$

(iii) Unused finance – 0; Unused craftsman time – 0

(iv) Too many chairs per table!

13 (i) Let x, y and z be the numbers of units of X, Y and Z produced respectively.
$\quad$ Maximise $3x + 2y + 5z$
$\quad$ Subject to $2x + 5y + 4z \leqslant 60$
$\qquad\qquad\quad 2y + z \leqslant 10$
$\qquad\qquad\quad 2x + 4y + 2z \leqslant 70$
$\qquad\qquad\quad 4x + 3y + 2z \leqslant 180$

(ii)

P	x	y	z	s_1	s_2	s_3	s_4	RHS
1	−3	−2	−5	0	0	0	0	0
0	2	5	4	1	0	0	0	60
0	0	2	1	0	1	0	0	10
0	2	4	2	0	0	1	0	70
0	4	3	2	0	0	0	1	180
1	−3	8	0	0	5	0	0	50
0	2	−3	0	1	−4	0	0	20
0	0	2	1	0	1	0	0	10
0	2	0	0	0	−2	1	0	50
0	4	−1	0	0	−2	0	1	160
1	0	3.5	0	1.5	−1	0	0	80
0	1	−1.5	0	0.5	−2	0	0	10
0	0	2	1	0	1	0	0	10
0	0	3	0	−1	2	1	0	30
0	0	5	0	−2	6	0	1	120
1	0	5.5	1	1.5	0	0	0	90
0	1	2.5	2	0.5	0	0	0	30
0	0	2	1	0	1	0	0	10
0	0	−1	−2	−1	0	1	0	10
0	0	−7	−6	−2	0	0	1	60

$\quad$ Solution: $x = 30, y = 0, z = 0, P = 90$.

(iii)

P	x	y	z	s_1	s_2	s_3	s_4	s_5	a_5	RHS
0	0	1	0	0	0	0	0	-1	1	5

(iv)

Q	P	x	y	z	s_1	s_2	s_3	s_4	s_5	a_5	RHS
1	0	0	1	0	0	0	0	0	-1	0	5
0	1	-3	-2	-5	0	0	0	0	0	0	0
0	0	0	1	0	0	0	0	0	-1	1	5
0	0	2	5	4	1	0	0	0	0	0	60
0	0	0	2	1	0	1	0	0	0	0	10
0	0	2	4	2	0	0	1	0	0	0	70
0	0	4	3	2	0	0	0	1	0	0	180

Start by minimising Q, pivoting on the y column. Drop the Q row and the a_5 column when Q attains 0, and proceed as usual.

Or use the single objective

P	x	y	z	s_1	s_2	s_3	s_4	s_5	a_5	RHS
1	-3	$-2-M$	-5	0	0	0	0	M	0	$-5M$

14 (i) Minimise $4x + 10y + 8z$

Subject to $x \geqslant 4(y + z)$

$\qquad\qquad y \geqslant 100$

$\qquad\qquad z \geqslant 80$

(ii) The first constraint has been written as $-x + 4y + 4z \leqslant 0$ using slack s_1.

The second constraint is $y \geqslant 100$ using surplus s_2 and artificial variable a_2.

The third constraint is $z \geqslant 80$ using surplus s_3 and artificial variable a_3.

Q is the sum of the artificial variables, $a_2 + a_3$.

The Q row has been rewritten using $y - s_2 + a_2 = 100$ and $z - s_3 + a_3 = 80$.

C is the cost, $4x + 10y + 8z$.

The final tableau shows $x = 720$, $y = 100$ and $z = 80$, with cost = £4520.

This is feasible since $Q = 0$. (It is also optimal since there are no non-artificial positive numbers in the C row.)

(iii) $P = 4x + 5y + 4z$

(iv)

P	x	y	z	s_1	s_2	s_3	s_4	RHS
1	0	0	0	0	5	4	1	4180
0	1	0	0	0	2.5	2	0.25	840
0	0	1	0	0	-1	0	0	100
0	0	0	1	0	0	-1	0	80
0	0	0	0	1	6.5	6	0.25	120

Produce 840 chairs, 100 round tables and 80 square tables.

15 (i) Let x, y and z be the numbers of litres of X, Y and Z respectively.

Row 1 ⟺ total profit $P = 10x + 10y + 20z$ (since 40p − 30p = 10p, etc.)

Row 2 ⟺ $5x + 2y + 10z \leqslant 10\,000$ (A's availability)

Row 3 ⟺ $2x + 4y + 5z \leqslant 12\,000$ (B's availability)

Row 4 ⟺ $8x + 3y + 5z \leqslant 8000$ (C's availability)

(ii)

P	x	y	z	s_1	s_2	s_3	RHS
1	−10	−10	−20	0	0	0	0
0	5	2	10	1	0	0	10 000
0	2	4	5	0	1	0	12 000
0	8	3	5	0	0	1	8 000
1	0	−6	0	2	0	0	20 000
0	0.5	0.2	1	0.1	0	0	1 000
0	−0.5	3	0	−0.5	1	0	7 000
0	5.5	2	0	−0.5	0	1	3 000
1	16.5	0	0	0.5	0	3	29 000
0	−0.05	0	1	0.15	0	−0.1	700
0	−8.75	0	0	0.25	1	−1.5	2 500
0	2.75	1	0	−0.25	0	0.5	1 500

Make 1500 litres of Y and 700 litres of Z, giving a profit of £290.

(iii) By 16.5p per litre

(iv)

Q	P	x	y	z	s_1	s_2	s_3	s_4	a	RHS
1	0	1	0	0	0	0	0	−1	0	500
0	1	−10	−10	−20	0	0	0	0	0	0
0	0	5	2	10	1	0	0	0	0	10 000
0	0	2	4	5	0	1	0	0	0	12 000
0	0	8	3	5	0	0	1	0	0	8 000
0	0	1	0	0	0	0	0	−1	1	500

Start by minimising Q until 0 (if feasible). Then drop Q and a and proceed to optimum.

16 (i) At $(0, 4.5)$, $P = 2.25$; at $(1, 4)$, $P = \frac{7}{3}$; at $(4, 2)$, $P = \frac{7}{3}$, and at $(5, 0)$, $P = \frac{5}{3}$.

So $P = \frac{7}{3}$ at either $(1, 4)$ or $(4, 2)$ (or anywhere on the line segment joining them).

(ii)

P	x	y	s_1	s_2	s_3	RHS
1	$-\frac{1}{3}$	$-\frac{1}{2}$	0	0	0	0
0	1	2	1	0	0	9
0	2	3	0	1	0	14
0	2	1	0	0	1	10
1	0	$-\frac{1}{3}$	0	0	$\frac{1}{6}$	$\frac{5}{3}$
0	0	$\frac{3}{2}$	1	0	$-\frac{1}{2}$	4
0	0	2	0	1	−1	4
0	1	$\frac{1}{2}$	0	0	$\frac{1}{2}$	5
1	0	0	0	$\frac{1}{6}$	0	$\frac{7}{3}$
0	0	0	1	$-\frac{3}{4}$	$\frac{1}{4}$	1
0	0	1	0	$\frac{1}{2}$	$-\frac{1}{2}$	2
0	1	0	0	$-\frac{1}{4}$	$\frac{3}{4}$	4

The solution is $P = \frac{7}{3}$ at $(4, 2)$.

(iii)

P	x	y	s_1	s_2	s_3	RHS
1	0	0	0	$\frac{1}{6}$	0	$\frac{7}{3}$
0	0	0	4	-3	1	4
0	0	1	2	-1	0	4
0	1	0	-3	2	0	1

This represents the other optimal vertex, $(1, 4)$.

(iv) The tableau is optimal at $(0, 0, 4)$ with value $\frac{7}{3}$.

Other solutions exist since there is (are) a non-basic variable(s) with zero in the objective row. $(1, 4, 0)$ and $(4, 2, 0)$

Chapter 2

❷ (Page 42)

1 A E G D – 12

2 A B – 8
A B C – 11
A E – 3
A E F – 4
A E G or A E F G – 10

3 There are $^7C_2 = \dfrac{7 \times 6}{2 \times 1} = 21$ altogether, so that is 15 more after considering all those involving A.

❷ (Page 43)

The number of ways of choosing two vertices from five is called $^5C_2$ (read as 'five choose 2'). The formula is

$$^5C_2 = \frac{5 \times 4}{2 \times 1} = 10$$ so there are ten ways of choosing two vertices from five, and therefore that many paths.

$$^{10}C_2 = \frac{10 \times 9}{2 \times 1} = 45$$

$$^nC_2 = \frac{n \times (n - 1)}{2 \times 1}$$

❷ (Page 48)

Shortest distance = 7; shortest path = $3 \to 2 \to 5 \to 4$

Exercise 2A (Page 48)

1 (i)

D(0)

	1	2	3	4
1	∞	9	∞	3
2	9	∞	2	∞
3	∞	2	∞	2
4	3	∞	2	∞

R(0)

	1	2	3	4
1	1	2	3	4
2	1	2	3	4
3	1	2	3	4
4	1	2	3	4

D(1)

	1	2	3	4
1	∞	9	∞	3
2	9	18	2	12
3	∞	2	∞	2
4	3	12	2	6

R(1)

	1	2	3	4
1	1	2	3	4
2	1	1	3	1
3	1	2	3	4
4	1	1	3	1

D(2)

	1	2	3	4
1	18	9	11	3
2	9	18	2	12
3	11	2	4	2
4	3	12	2	6

R(2)

	1	2	3	4
1	2	2	2	4
2	1	1	3	1
3	2	2	2	4
4	1	1	3	1

D(3)

	1	2	3	4
1	18	9	11	3
2	9	4	2	4
3	11	2	4	2
4	3	4	2	4

R(3)

	1	2	3	4
1	2	2	2	4
2	1	3	3	3
3	2	2	2	4
4	1	3	3	3

D(4)

	1	2	3	4
1	6	7	5	3
2	7	4	2	4
3	5	2	4	2
4	3	4	2	4

R(4)

	1	2	3	4
1	4	4	4	4
2	3	3	3	3
3	4	2	2	4
4	1	3	3	3

1 (ii) Final matrices:

D(8)

	1	2	3	4	5	6	7	8
1	6	7	5	3	9	13	8	18
2	7	4	2	4	6	10	8	15
3	5	2	4	2	4	8	6	13
4	3	4	2	4	6	10	5	15
5	9	6	4	6	8	10	10	9
6	13	10	8	10	10	14	7	12
7	8	8	6	5	10	7	10	10
8	18	15	13	15	9	12	10	18

R(8)

	1	2	3	4	5	6	7	8
1	4	4	4	4	4	4	4	4
2	3	3	3	3	3	3	3	3
3	4	2	2	4	5	6	7	5
4	1	3	3	3	3	3	7	3
5	3	3	3	3	3	6	3	8
6	3	3	3	3	5	7	7	8
7	4	3	3	4	3	6	4	8
8	5	5	5	5	5	6	7	5

2 (i)

D(0)

	1	2	3	4
1	∞	7	9	∞
2	7	∞	18	6
3	9	18	∞	3
4	∞	6	3	∞

R(0)

	1	2	3	4
1	1	2	3	4
2	1	2	3	4
3	1	2	3	4
4	1	2	3	4

D(1)

	1	2	3	4
1	∞	7	9	∞
2	7	14	16	6
3	9	16	18	3
4	∞	6	3	∞

R(1)

	1	2	3	4
1	1	2	3	4
2	1	1	1	4
3	1	1	1	4
4	1	2	3	4

D(2)

	1	2	3	4
1	14	7	9	13
2	7	14	16	6
3	9	16	18	3
4	13	6	3	12

R(2)

	1	2	3	4
1	2	2	3	2
2	1	1	1	4
3	1	1	1	4
4	2	2	3	2

D(3)

	1	2	3	4
1	14	7	9	12
2	7	14	16	6
3	9	16	18	3
4	12	6	3	6

R(3)

	1	2	3	4
1	2	2	3	3
2	1	1	1	4
3	1	1	1	4
4	3	2	3	3

D(4)

	1	2	3	4
1	14	7	9	12
2	7	12	9	6
3	9	9	6	3
4	12	6	3	6

R(4)

	1	2	3	4
1	2	2	3	3
2	1	4	4	4
3	1	4	4	4
4	3	2	3	3

(ii)

D(0)

	1	2	3	4	5
1	∞	7	9	∞	∞
2	7	∞	18	6	∞
3	9	18	∞	3	1
4	∞	6	3	∞	1
5	∞	∞	1	1	∞

R(0)

	1	2	3	4	5
1	1	2	3	4	5
2	1	2	3	4	5
3	1	2	3	4	5
4	1	2	3	4	5
5	1	2	3	4	5

D(1)

	1	2	3	4	5
1	∞	7	9	∞	∞
2	7	14	16	6	∞
3	9	16	18	3	1
4	∞	6	3	∞	1
5	∞	∞	1	1	∞

R(1)

	1	2	3	4	5
1	1	2	3	4	5
2	1	1	1	4	5
3	1	1	1	4	5
4	1	2	3	4	5
5	1	2	3	4	5

D(2)

	1	2	3	4	5
1	14	7	9	13	∞
2	7	14	16	6	∞
3	9	16	18	3	1
4	13	6	3	12	1
5	∞	∞	1	1	∞

R(2)

	1	2	3	4	5
1	2	2	3	2	5
2	1	1	1	4	5
3	1	1	1	4	5
4	2	2	3	2	5
5	1	2	3	4	5

D(3)

	1	2	3	4	5
1	14	7	9	12	10
2	7	14	16	6	17
3	9	16	18	3	1
4	12	6	3	6	1
5	10	17	1	1	2

R(3)

	1	2	3	4	5
1	2	2	3	3	3
2	1	1	1	4	1
3	1	1	1	4	5
4	3	2	3	3	5
5	3	3	3	4	3

D(4)

	1	2	3	4	5
1	14	7	9	12	10
2	7	12	9	6	7
3	9	9	6	3	1
4	12	6	3	6	1
5	10	7	1	1	2

R(4)

	1	2	3	4	5
1	2	2	3	3	3
2	1	4	4	4	4
3	1	4	4	4	5
4	3	2	3	3	5
5	3	4	3	4	3

D(5)

	1	2	3	4	5
1	14	7	9	11	10
2	7	12	8	6	7
3	9	8	2	2	1
4	11	6	2	2	1
5	10	7	1	1	2

R(5)

	1	2	3	4	5
1	2	2	3	3	3
2	1	4	4	4	4
3	1	5	5	5	5
4	5	2	5	5	5
5	3	4	3	4	3

(iii) Final matrices:

D(7)

	1	2	3	4	5	6	7
1	14	7	9	17	12	14	17
2	7	10	9	10	5	7	10
3	9	9	8	9	4	6	9
4	17	10	9	6	5	3	6
5	12	5	4	5	4	2	5
6	14	7	6	3	2	4	3
7	17	10	9	6	5	3	6

R(7)

	1	2	3	4	5	6	7
1	2	2	3	2	2	2	2
2	1	5	5	5	5	5	5
3	1	5	5	5	5	5	5
4	6	6	6	6	6	6	6
5	2	2	3	6	6	6	6
6	5	5	5	4	5	5	7
7	6	6	6	6	6	6	6

❷ (Page 50)

The pop tour could be shortened by using the route London → Brighton → Bristol → Birmingham → Manchester → Liverpool → Glasgow → Aberdeen → Newcastle → Sheffield → Nottingham → Oxford → London. In fact this reduces the distance from about 1590 miles to about 1360 miles.

Exercise 2B (Page 50)

The optimal tour is: Worcester → Evesham → Tewkesbury → Cheltenham → Gloucester → Ross → Hereford → Malvern → Worcester; 104 miles

❷ (Page 56)

One possible counter-example is shown in figure 2.16.

❷ (Page 58)

No, 101 is the best lower bound, achievable in several ways.

Exercise 2C (Page 63)

1 Upper bound (using nearest neighbour starting from A) 41, lower bound (by deleting A or B or D) 40

2 Upper bound (using nearest neighbour starting from Dover) 1258

Lower bounds: Town deleted

London	793
Chester	975
Dover	793
Glasgow	**1022**
Oxford	878
Plymouth	880

(The optimal tour = London → Plymouth → Glasgow → Chester → Oxford → Dover → London; 1256)

3 Weston → Burnham → Bridgwater → Glastonbury → Wells → Bath → Cheddar → Weston; 95

4 Birmingham → Gloucester → Hereford → Shrewsbury → Stoke → Sheffield → Nottingham → Northampton → Birmingham; 357

5 St. Hélier → Corbiere → Devil's Hole → Trinity Church → Rozel → Mount Orgueil → La Hogue Bie → St. Hélier → La Rocque Point→ St. Hélier; 37.5

6 Optimal tour = A → B → C → F → E → H → G → D → A; 136

7 (i) Strawberry → orange → lemon → lime → raspberry → strawberry; 73 minutes
(ii) No; this gives 77 minutes

8 Optimum distance 378

9 A_1 can be added to the table with distance to and from A of ∞, and with distances to and from B, C, D and E equal to the distances from A to the towns. The solution to the travelling salesperson problem will then be of the form A ***A_1*** A, and this can be interpreted as two separate Hamilton cycles, one for each lorry. The two tours are A → B → C → A and A → D → E → A; total length 237

Exercise 2D (Page 69)

1 (i) 201 + 15   **(ii)** 107 + 34

2 (i) 170 000 + 2900   **(ii)** 170 000 + 2800

3 274 + 33

4 83 + 18; e.g. A → B → A → D → I → H → F → D → H → G → F → C → B → F → C → E → G → H → I → A

5 101 + 23; e.g. St Oz → A → St Oz → G → F → E → F → D → B → C → B → A → D → G → St Oz → I → G → H → I → H → St Oz

6 It is simplest to label the nodes and describe the route by giving a list of nodes. The diagram shows the map relabelled in this way. The heavy lines indicate those arcs to be traversed twice.

One possible route is G.P.O. → A → B → M → L → K → J → I → H → G → F → E → K → P → Q → P → J → F → E → D → C → B → M → C → D → L → O → U → V → N → M → N → O → Q → R → I → H → T → G → T → S → R → S → U → V → W → A → W → G.P.O.

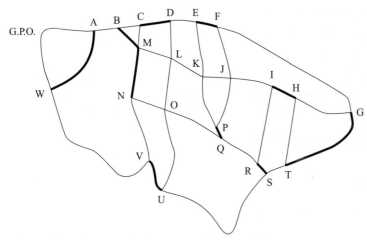

Exercise 2E (Page 73)

1 (i) 3   **(ii)** $5 \times 3 = 15$ **(iii)** $7 \times 5 \times 3 = 105$ **(iv)** 654 729 075

2 (i) B, D, F, G

(ii) Eulerian graphs have no odd nodes.

(iii) BD/FG = 550, BF/DG = 250, BG/DF = 550

e.g. A → B → F → B → C → D → E → G → E → D → G → F → E → C → A, Length = 1950 + 250 = 2200

3 (i) Order = BECDF

Arcs = BE, BC, ED, DF, Length = 37

(ii) 19 + 37 = 56

(iii) Deleting A and its arcs deletes two arcs from the optimal Hamilton cycle. The remainder of that cycle is a connector for the remainder of the network. Thus the weight of a minimum connector for the remainder, plus the weights of the two least weight deleted arcs, cannot be of greater weight than the optimal Hamilton cycle.

(iv) Route = A → E → B → C → D → F → A, Length = 59, quite close to the lower bound.

4 (i) (a) A, C, F, I; need to be connected in pairs by repeated arcs to give a route which traverses all arcs.

(b) AC/FI = 320m, AF/CI = 680m, AI/CF = 680m

2300 + 320 = 2620

(c)

Intersection	B	C	D	E	F	G	H	I
Number of visits	3	2	2	2	3	2	2	2

(ii) (a) Repeat each arc.

(b) All nodes now even so traversable. Distance = 4600

(iii) This needs repeated arcs as in part (ii), but they are directed, each pair being directed in opposite directions. However, the resulting network is traversable.

5 (i) **(a)** All of order 2.

(b) All nodes even so the network is Eulerian and therefore traversable.

Single rope needs only to be as long as the sum of the four section lengths.

(ii) **(a)**

Node	A	B	C	D	E	F
Order	3	3	1	1	2	2

In traversing a network vertices are encountered and left along different arcs. So there needs to be an even number of arcs incident upon each vertex.

(b) Some sections will have to have rope along them twice, so the total length required will exceed the sum of the lengths of the six sections.

(c) Need to have AD and BC repeated, e.g. $C \to B \to F \to A \to D \to A \to E \to B \to C$

(d) $C \to B \to F \to A \to D \to A \to E \to B$ or $C \to B \to E \to A \to D \to A \to F \to B$ or
$C \to B \to X \to A \to Y \to B \to Z \to A \to D$, where {X, Y, Z} is any permutation of {E, E, F} or {E, F, F}.
(There are six such possibilities.)

(iii) Start at C and end at D or vice-versa; e.g. $C \to B \to F \to A \to E \to B \to A \to D$

6 (i) **(b)**

K	L	M	N	O	P	Q	R	S
50	59	67	74	69	75	78	89	90

(ii) $J \to K \to L \to M \to N \to Q \to P \to O \to N \to Q \to R \to S \to Q \to N \to O \to K \to J$, Length = 221 km

(iii) Pair odd nodes and connect by repeating shortest distances. Best pairing JM/OS.
Route = e.g. $J \to K \to L \to M \to K \to O \to N \to Q \to \underline{S \to Q \to N \to O} \to P \to S \to R \to Q \to P \to N \to M \to K \to L \to J$, Length = 298 km

7 (i) $A \to B \to D \to E \to C \to F \to A$ with cost 640
Another tour is $A \to B \to D \to E \to F \to C \to A$ with cost 537

(ii) 5! = 120

(iii)

From \ To	A	B	C	D	E	F
A	–	65	80	78	110	165
B	75	–	97	55	113	130
C	80	90	–	70	90	340
D	90	65	90	–	75	250
E	110	90	80	45	–	82
F	165	130	320	195	100	–

(iv) e.g. B to E. Without taxes cheapest is BDE, costing 60. With taxes cheapest is BE costing 113.

(v) No difference. All taxes are incurred once and only once giving an increase of 140 on all tours.

8 (ii) $P \to V \to U$, Cost = 110 (£1.10)

(iii) P and U are odd. $P \to V$ and $V \to U$ have to be repeated.
Route = e.g. $V \to P \to Q \to U \to T \to V \to S \to T \to R \to Q \to S \to P \to V \to U \to V$, Cost = 809 (£8.09)

(iv) PR/SU = 148, PS/RU = 132, PU/RS = 145
Route = e.g. $P \to Q \to U \to T \to V \to S \to T \to R \to Q \to S \to P \to V \to U \to T \to R \to S \to P$, Cost = 866 (£8.66)

9 (i) e.g. $1 \to 5 \to 1 \to 3 \to 5 \to 4 \to 3 \to 2 \to 1$, Length = 23 + 2 = 25

(ii) **(a)**

	1	2	3	4	5
1	2	4	1	7	2
2	4	8	5	11	6
3	1	5	2	6	2
4	7	11	6	12	3
5	2	6	2	3	4

	1	2	3	4	5
1	3	2	3	3	5
2	1	1	3	3	1
3	1	2	1	4	5
4	3	3	3	3	5
5	1	1	3	4	1

(b) $4 \to 5 \to 1 \to 2$

(c)

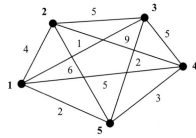

(iii) $(5 + 2 + 3) + (1 + 2) = 13$

(iv) $1 \to 3 \to 5 \to 4 \to 2 \to 1$, Length $= 19$

(v) $1 \to 3 \to 5 \to 4 \to 5 \to 1 \to 2 \to 1$

10 (i) The colour is equivalent to the town, the cost of cleaning to the cost of travel, and using a colour to visiting a town.

(ii) WYGBRW – 13

YGBRWY – 13

BGYRWB – 17 or BGYWRB – 13 or BGRYWB – 14

GBRYWG – 12

RBGYWR – 13 or RGBYWR – 15

(iii) GBRYWG = WGBRYW

(iv) Connectedness is not defined in a digraph.

11 (i)

	1	2	3	4	5
1	∞	5	∞	2	7
2	5	∞	4	2	∞
3	∞	4	∞	1	1
4	2	2	1	∞	3
5	7	∞	1	3	∞

	1	2	3	4	5
1	1	2	3	4	5
2	1	2	3	4	5
3	1	2	3	4	5
4	1	2	3	4	5
5	1	2	3	4	5

(ii)

	1	2	3	4	5
1	10	5	9	2	7
2	5	8	4	2	5
3	9	4	8	1	1
4	2	2	1	2	2
5	7	5	1	2	2

	1	2	3	4	5
1	2	2	2	4	5
2	1	3	3	4	3
3	2	2	2	4	5
4	1	2	3	3	3
5	1	3	3	3	3

(iii) $5 \to 3 \to 4 \to 2$, 4 hours

(iv) $1 + 2 + 7$ (minimum connector) $= 10$ hours

(v) $1 \to 4 \to 3 \to 5 \to 2 \to 1$, 12 hours

(vi) $4 \to 3 \to 5 \to 3 \to 4 \to 2 \to 4 \to 1 \to 4$

(vii) $25 + 4 + 1$ (best pairing) $= 30$ hours

12 (i) The entry in row E and column B of the distance matrix is 11 so the shortest distance is 11 km.

The entry in row E and column B of the route matrix is D, so the route starts ED …

The entry in row D and column B of the route matrix is B, so the route is EDB.

(ii) $A \to F \to B \to C \to D \to E \to A$, Length 32 km, AFABCDEA

(iii) $A \to B \to C \to D \to E \to F \to A$, Length 29.5 km

(iv) The tracks are not defined by beaches: we need to have nodes for track junctions.

Either $6 + 5 = 11$ or $3 + 5 = 8$ (the latter by cutting out the 2-nodes at B, D and F)

(v) Need to repeat $1 + 1.5 + 2.5 + 1 + 1.5 = 7.5$ km, e.g. $A \to C \to B \to A \to F \to D \to E \to A$,

Distance $= 32 + 7.5 = 39.5$ km

Chapter 3

Exercise 3A (Page 91)

1 (i) 0

(ii) $-£\frac{1}{37}$ or -2.70p (3sf)

2 (i) $-£\frac{21}{36} \approx -58$p

(ii) £22

(iii) At least $\sqrt[3]{8 \times 12^4} - 2 \approx 52.95$

3 (i) $\frac{125}{216}, \frac{75}{216}, \frac{15}{216}$

(ii) $-£\frac{17}{216} \approx -7.9$p

4 (i)   Insure, fine = £3500

           Insure, rain = −£5500

      So, insure = £1700

           Don't insure, fine = £5000

           Don't insure, rain = −£10 000

      So, don't insure = £2000

      Best policy is not to take out the insurance.

(ii)   Insure, fine = £3500

           Insure, rain = −£5500

      So, insure = £500

           Don't insure, fine = £5000

           Don't insure, rain = −£10 000

      So, don't insure = £0

      No, would have taken out the insurance.

5 EMV if die is not rolled again after a 6 is £3.50.

   EMV if it is rolled again when possible is £3.75.

   So optimal policy is to roll again if possible.

6 (i)   Don't test and launch = £95 000

           Don't test and sell = £90 000

      So don't test = £95 000

               Test, favourable, launch = £165 000

               Test, favourable, sell = £85 000

           So test, favourable = £165 000

               Test, unfavourable, launch = £75 000

               Test, unfavourable, sell = £85 000

           So test, unfavourable = £85 000

      So test = £98 333

(ii) Test, launch if favourable and sell if not, EMV = £98 333.

(iii) £8333

7   Don't survey and drill = −£100 000

      Don't survey and don't drill = £0

     So don't survey = £0

           Survey, promising, drill = £650 000

           Survey, promising, don't drill = −£150 000

       So survey, promising = £650 000

           Survey, not promising, drill = −£550 000

           Survey, not promising, don't drill = −£150 000

       So survey, not promising = −£150 000

     So survey = £50 000

   So best course of action is survey, drill if promising and don't drill if not, EMV = £50 000.

8 (i) Refuse (EMV = –£0.025 million)

(ii) Don't survey and drill = –£25 000
Don't survey and don't drill = £0
So don't survey = £0
    Survey, promising, drill = £100 000
    Survey, promising, don't drill = –£200 000
    So survey, promising = £100 000
    Survey, not promising, drill = –£550 000
    Survey, not promising, don't drill = –£200 000
    So survey, not promising = –£200 000
So survey = –£50 000
So best course of action is to abandon the project.

(iii) £150 000

9 (i) Don't take 2nd bet = £100 000
    Take 2nd bet, win = £1 500 000
    Take 2nd bet, lose = £0
So take 2nd bet = £150 000
So optimal decision is to take the 2nd bet.

(ii) Utility of not taking the bet is $\sqrt{100\,000} \approx 316.2$
Utility of taking bet is $0.1 \times \sqrt{1\,500\,000} \approx 122.5$
So don't take the bet.

Exercise 3B (Page 94)

1 (i) Final EMV = £1600, by insuring (EMV = £1000 if no insurance)

(ii) EMVs
    Don't consult and insure = £1600
    Don't consult and don't insure = £1000
So don't consult = £1600
    Consult, rain forecast, insure = £1000
    Consult, rain forecast, don't insure = –£2000
    So consult, rain forecast = £1000
    Consult, rain not forecast, insure = £1800
    Consult, rain not forecast, don't insure = £2000
    So consult, rain not forecast = £2000
So consult = £1750

(iii) Consultation worth £150

2 (i) Expected cost of store first = £3382m, so build now at a cost advantage of £882m.

(ii) Defer at a cost advantage of £(2500 – 2080)m = £420m

(iii)

3 (i) (a)

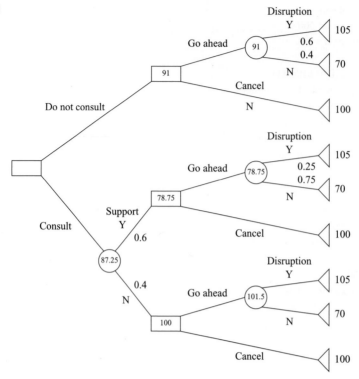

(b) Value of consultation = 3.75 billion pesetas

(ii) (a) $0.8 \times 105 + 0.2 \times 70 = 98$

Value = 10.75 billion pesetas

(b) $100 - 87.25 = 12.75$

Minimum probability $= \frac{6}{7}$

4 (i) In order (£): 500, 500, 800, 0, 550, 550, 850, 50, 550, 550, 850, 50

(iii) Expected costs

No further test, treat = 500

No further test, don't treat = 640

So no further test = 500

Further test, positive, treat = 550

Further test, positive, don't treat = 810

So further test, positive = 550

Further test, negative, treat = 550

Further test, negative, don't treat = 530

So further test, negative = 530

So further test = 541

(iv) Don't send for further test but send for treatment: expected cost = £500.

5 (i) Require $100p + 20(1 - p) = 50$, giving $p = \frac{3}{8}$.

(ii) (a)

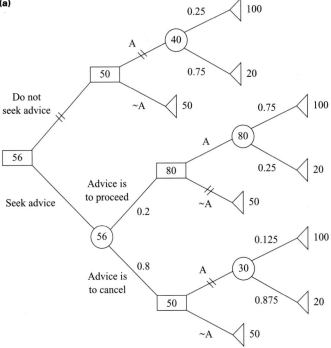

(b) Value of advice = 6

(c) Assuming the payoff is not more than 80, require $X = 16 + 0.8X$, giving $X = 80$.

6 (i)

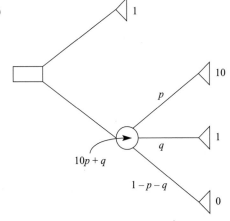

(ii) 0.05

(iii), (iv)

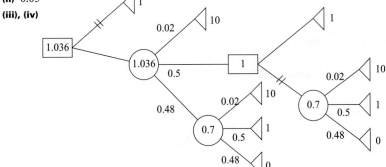

Go again, then accept the £1 if it is offered again.

7 (i) £1150, £900, £1050, £1030

(ii)

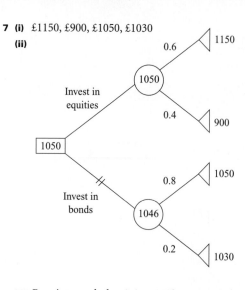

(iii) Require p such that $0.6 \times 1150^p + 0.4 \times 900^p = 0.8 \times 1050^p + 0.2 \times 1030^p$
Letting $f(p) = 0.6 \times 1150^p + 0.4 \times 900^p - 0.8 \times 1050^p - 0.2 \times 1030^p$
$\quad f(0.5) = 0.0053$
$\quad f(0.45) = -0.0003$

8 (i)

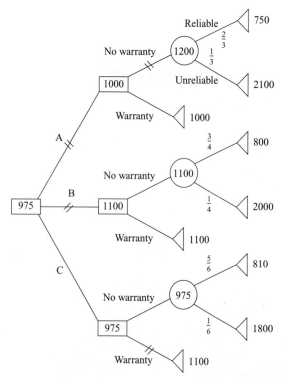

Choose a car of type C and do not buy a warranty; EMV = £975

(ii)

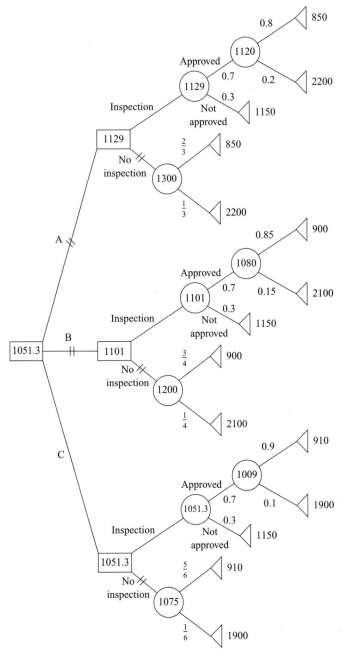

Choose a second-hand car of type C, and have it inspected; EMV = £1051.30

(iii) A £171, B £99, C £23.70

(iv) Require x such that $0.7 \times 1009 + 0.3x = 1075$, giving $x = 1229$.

Chapter 4

❷ (Page 103)

Ask of one sentinel 'If I ask the other sentinel which is the way to salvation, what will be the answer?'
This question 'collects' one lie, so go the other way.

❷ (Page 103)

Ask of one sentinel about another sentinel 'If I repeatedly ask him which is the way to salvation will left or right dominate his answers?'

If you are asking the random responder then the answer you will get will either be 'left' or 'right'.

If you are asking *about* the random responder then the answer you will get will either be 'yes' or 'no'! (You were warned!) You can then eliminate him, reducing the problem to the above case. If you get the answer 'left' or the answer 'right', then the sentinel you are asking *about* will either be the liar or the truth teller. Now ask *him* the same question about the third sentinel. If the answer is 'left' or 'right', then go the other way. If the answer is 'yes' or 'no', then take the direction opposite to that given in answer to your first question.

Exercise 4A (Page 107)

1 $m \Rightarrow p$

2 $\sim n \Rightarrow p$

3 $q \Rightarrow \sim n$

4 $q \Leftrightarrow (n \wedge \sim m)$

5 $q \Rightarrow \sim p$

6 *If you don't have an umbrella with you then the weather is fine.*

7 *If the weather is not fine then it is wet.*

8 *If it is warm and not wet then it is fine.*

9 *If it is wet then you have an umbrella with you.*

10 *You have an umbrella and it is wet if and only if the weather is not fine.*

Exercise 4B (Page 109)

1 a	$\vee$	$\sim a$
0	1	1
1	1	0

2 $\sim$	$(a$	$\vee$	$b)$	$\Leftrightarrow$	$\sim$	a	$\wedge$	$\sim$	b
1	0	0	0	1	1	0	1	1	0
0	0	1	1	1	1	0	0	0	1
0	1	1	0	1	0	1	0	1	0
0	1	1	1	1	0	1	0	0	1

3 $(a$	$\Rightarrow$	$b)$	$\Leftrightarrow$	$(\sim$	b	$\Rightarrow$	$\sim$	$a)$
0	1	0	1	1	0	1	1	0
0	1	1	1	0	1	1	1	0
1	0	0	1	1	0	0	0	1
1	1	1	1	0	1	1	0	1

4

[a	∧	(a	⇒	b)]	⇒	b
0	0	0	1	0	1	0
0	0	0	1	1	1	1
1	0	1	0	0	1	0
1	1	1	1	1	1	1

5

a	∧	(b	∨	c)	⇔	(a	∧	b)	∨	(a	∧	c)
0	0	0	0	0	1	0	0	0	0	0	0	0
0	0	0	1	1	1	0	0	0	0	0	0	1
0	0	1	1	0	1	0	0	1	0	0	0	0
0	0	1	1	1	1	0	0	1	0	0	0	1
1	0	0	0	0	1	1	0	0	0	1	0	0
1	1	0	1	1	1	1	0	0	1	1	1	1
1	1	1	1	0	1	1	1	1	1	1	0	0
1	1	1	1	1	1	1	1	1	1	1	1	1

6

[p	∨	(~	p	∨	q)]	∨	(~	p	∧	~	q)
0	1	1	0	1	0	1	1	0	1	1	0
0	1	1	0	1	1	1	1	0	0	0	1
1	1	0	1	0	0	1	0	1	0	1	0
1	1	0	1	1	1	1	0	1	0	0	1

7

[(u	⇒	~	m)	∧	(e	⇒	m)]	⇒	(e	⇒	~	u)
0	1	1	0	1	0	1	0	1	0	1	1	0
0	1	1	0	0	1	0	0	1	1	1	1	0
0	1	0	1	1	0	1	1	1	0	1	1	0
0	1	0	1	1	1	1	1	1	1	1	1	0
1	1	1	0	1	0	1	0	1	0	1	0	1
1	1	1	0	0	1	0	0	1	1	0	0	1
1	0	0	1	0	0	1	1	1	0	1	0	1
1	0	0	1	0	1	1	1	1	1	0	0	1

The argument is valid.

8 Let b be the proposition *The bowler has bowled a bouncer.*

Let i be the proposition *The batsman is injured.*

(i) You are given $b \Rightarrow i$ and $\sim i$.

But $(b \Rightarrow i) \Leftrightarrow (\sim i \Rightarrow \sim b)$.

So you have $\sim i$ and $(\sim i \Rightarrow \sim b)$, giving $\sim b$ (from question 4).

(ii) Nothing can be deduced from i. He might have fallen over.

9 (i) If the probability of a head is not $\frac{1}{2}$ then it is not a fair coin.

(ii) x and y are integers. If xy is odd then x is odd.

(iii) Triangles which are not similar are not congruent.

(iv) If you don't remember Long John Silver then you have not read *Treasure Island*.

Remember, you are not concerned with the absolute truth of the propositions (you might have seen the film), but rather with their logical equivalence.

10 (i) Let c be the proposition *Alice has not eaten her cake.*

Then the compound proposition is $\sim(c \wedge \sim c)$.

~	(c	∧	~	c)
1	0	0	1	0
1	1	0	0	1

(ii) Let n be the proposition $x^2 = 9$.

Let tp be the proposition $x = 3$.

Let tn be the proposition $x = -3$.

Then the compound proposition is $\{[n \Rightarrow (tp \vee tn)] \wedge \sim tp\} \Rightarrow \sim n$

{[n	⇒	(tp	∨	tn)]	∧	~	tp	⇒	~	n
0	1	0	0	0	1	1	0	1	1	0
0	1	0	1	1	1	1	0	1	1	0
0	1	1	1	0	0	0	1	1	1	0
0	1	1	1	1	0	0	1	1	1	0
1	0	0	0	0	0	1	0	1	0	1
1	1	0	1	1	1	1	0	0	0	1
1	1	1	1	0	0	0	1	1	0	1
1	1	1	1	1	0	0	1	1	0	1

The argument is not valid, as can be seen by considering the case in which n is true, tp is false and tn is true. (You might argue that two lines of the table should be deleted since you cannot have both tp and tn true at the same time. But that is not deducible from what is given, nor is it relevant to the logic.)

(iii) Let e^2 be the proposition *Integer x^2 is even*.

Let e^1 be the proposition *Integer x is even*.

Then the compound proposition is $(\sim e^1 \Rightarrow \sim e^2) \Rightarrow (e^2 \Rightarrow e^1)$.

The truth table for $(\sim e^1 \Rightarrow \sim e^2) \Leftrightarrow (e^2 \Rightarrow e^1)$ can be found in the answer to question 3.

11 (i) $[(p \vee q) \wedge \sim q] \Rightarrow p$

[(p	∨	q)	∧	~	q]	⇒	p
0	0	0	0	1	0	1	0
0	1	1	0	0	1	1	0
1	1	0	1	1	0	1	1
1	1	1	0	0	1	1	1

e.g. Either the weather is fine or you have an umbrella. You do not have an umbrella. So the weather must be fine.

(ii) $[(p \Rightarrow q) \wedge p] \Rightarrow q$

[(p	⇒	q)	∧	p]	⇒	q
0	1	0	0	0	1	0
0	1	1	0	0	1	1
1	0	0	0	1	1	0
1	1	1	1	1	1	1

e.g. If you are honest you will be respected. You are honest. So you are respected.

(iii) $[(p \Rightarrow q) \wedge \sim q] \Rightarrow \sim p$

[(p	⇒	q)	∧	~	q]	⇒	~	p
0	1	0	1	1	0	1	1	0
0	1	1	0	0	1	1	1	0
1	0	0	0	1	0	1	0	1
1	1	1	0	0	1	1	0	1

e.g. If the weather is fine you have an umbrella. You do not have an umbrella. So the weather is not fine.

12 $\cos(300°) = \frac{1}{2}$

13 (i) Not equivalent. Part (b) says that either he runs but does not jump, or vice versa, or both. $(\sim(\sim r \wedge \sim j) \Leftrightarrow r \vee j)$. This is not the same as $r \wedge j$.

~	(~	r	∧	~	j)	⇔	r	∨	j
0	1	0	1	1	0	1	0	0	0
1	1	0	0	0	1	1	0	1	1
1	0	1	0	1	0	1	1	1	0
1	0	1	0	0	1	1	1	1	1

(ii) These are equivalent. $\sim(ph \wedge pr) \Leftrightarrow \sim ph \vee \sim pr$ (see part (i))

(iii) These are not equivalent. $\sim(\sim g \vee \sim t) \Leftrightarrow g \wedge t$ (see question 2)

So part (a) is equivalent to *He is guilty and telling the truth.*

14 The argument is a correct deduction. It shows that *if* $0 = 1$ then $1 = 1$. Furthermore, $1 = 1$ is true. But this does not imply that $0 = 1$. Let p be $0 = 1$ and q be $1 = 1$.

In truth table form:

$[(p$	$\Rightarrow$	$q)$	$\wedge$	$q]$	$\Rightarrow$	p
0	1	0	0	0	**1**	0
0	1	1	1	1	**0**	0
1	0	0	0	0	**1**	1
1	1	1	1	1	**1**	1

This shows that a problem occurs where p is false and q true. Then $p \Rightarrow q$ is true. This is the case confronting the student.

Exercise 4C (Page 114)

1 $a \wedge b \wedge (a \vee c) \wedge [b \vee (c \wedge a) \vee d]$

$= [a \wedge (a \vee c)] \wedge [b \wedge (b \vee (c \wedge a) \vee d)]$ (commutative rules)

$= a \wedge b$ (absorption rules)

2 (i) $(a \wedge b \wedge c) \vee (\sim a \wedge b \wedge c)$

$= [a \wedge (b \wedge c)] \vee [\sim a \wedge (b \wedge c)]$ (associative rules)

$= (a \vee \sim a) \wedge (b \wedge c)$ (distributive rules)

$= 1 \wedge (b \wedge c)$ (complement rules)

$= b \wedge c$ (identity rules)

(ii) $\sim[(p \wedge q) \vee \sim p]$

$= \sim[(p \vee \sim p) \wedge (q \vee \sim p)]$ (distributive rules)

$= \sim[1 \wedge (q \vee \sim p)]$ (complement rules)

$= \sim(q \vee \sim p)$ (absorption rules)

$= \sim q \wedge \sim(\sim p)$ (De Morgan's rules)

$= \sim q \wedge p$ (double negation)

3 (i) $[(p \Rightarrow q) \Rightarrow p] \Rightarrow p = \sim[\sim(\sim p \vee q) \vee p] \vee p$

$\qquad\qquad\qquad\qquad\qquad = \sim[(p \wedge \sim q) \vee p] \vee p$

$\qquad\qquad\qquad\qquad\qquad = \sim p \vee p$

$\qquad\qquad\qquad\qquad\qquad = 1$

(ii) LHS: $p \Rightarrow (q \Rightarrow r) \quad = \sim p \vee (\sim q \vee r)$

$\qquad\qquad\qquad\qquad\qquad = \sim p \vee \sim q \vee r$

$\quad$ RHS: $(p \wedge q) \Rightarrow r \quad = \sim(p \wedge q) \vee r$

$\qquad\qquad\qquad\qquad\qquad = \sim p \vee \sim q \vee r$

(iii) $[(a \vee b) \vee (a \Rightarrow c) \wedge (b \Rightarrow c)] \Rightarrow c$

$= \sim[(a \vee b) \wedge (\sim a \vee c) \wedge (\sim b \vee c)] \vee c$

$= \sim[(a \vee b) \wedge \{(\sim a \wedge \sim b) \vee c\}] \vee c$

$= \sim[(a \vee b) \wedge \{\sim(a \vee b) \vee c\}] \vee c$

$= \sim[\{(a \vee b) \wedge \sim(a \vee b)\} \vee \{(a \vee b) \wedge c\}] \vee c$

$= \sim[0 \vee \{(a \vee b) \wedge c\}] \vee c$

$= \sim[(a \vee b) \wedge c] \vee c$

$= \sim(a \vee b) \vee \sim c \vee c$

$= \sim(a \vee b) \vee 1$

$= 1$

4 (i) $(a \wedge b) \vee \sim a \vee \sim b = (a \wedge b) \vee \sim(a \wedge b) = 1$

(ii) $(a \vee b) \wedge \sim a \wedge \sim b = (a \vee b) \wedge \sim(a \wedge b) = 0$

(iii) $a \vee \sim[(\sim b \vee a) \wedge b] = a \vee \sim[(\sim b \wedge b) \vee (a \wedge b)] = a \vee \sim[0 \vee (a \wedge b)] = a \vee \sim(a \wedge b) = a \vee \sim a \vee \sim b = 1 \vee \sim b = 1$

(iv) $a \wedge \sim[(\sim b \wedge a) \vee b] = a \wedge \sim[(\sim b \vee b) \wedge (a \vee b)] = a \wedge \sim[1 \wedge (a \vee b)] = a \wedge \sim(a \vee b) = a \wedge \sim a \wedge b = 0 \wedge b = 0$

(v) $(a \vee b) \wedge (\sim a \vee \sim b) = [(a \vee b) \wedge \sim a] \vee [(a \vee b) \wedge \sim b] = (b \wedge \sim a) \vee (a \wedge \sim b) = (a \wedge \sim b) \vee (\sim a \wedge b)$

(vi) $a \wedge (\sim a \vee b) = (a \wedge \sim a) \vee (a \wedge b) = 0 \vee (a \wedge b) = a \wedge b$

(vii) $a \vee (\sim a \wedge b) = (a \vee \sim a) \wedge (a \vee b) = 1 \wedge (a \vee b) = a \vee b$

(viii) $(a \vee b \vee c) \wedge (a \vee b) = [(a \vee b) \wedge (a \vee b)] \vee [c \wedge (a \vee b)]$
$(a \vee b) \vee [c \wedge (a \vee b)] = a \vee b$ (absorption)

(ix) $(a \wedge b \wedge c) \vee (a \wedge b) = [(a \wedge b) \vee (a \wedge b)] \wedge [c \vee (a \wedge b)]$
$(a \wedge b) \wedge [c \vee (a \wedge b)] = a \wedge b$ (absorption)

(x) $(a \vee b) \wedge (a \vee \sim b) = a \vee (b \wedge \sim b) = a \vee 0 = a$

(xi) $(a \wedge b) \vee (a \wedge \sim b) = a \wedge (b \vee \sim b) = a \wedge 1 = a$

(xii) $a \vee [a \wedge (b \vee 1)] = a \vee (a \wedge b) = a$ (absorption)

(xiii) $a \wedge [a \vee (b \wedge 1)] = a \wedge (a \vee b) = a$ (absorption)

Exercise 4D (Page 119)

1 (i)   **(ii)**

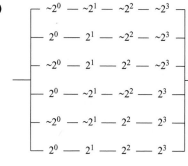

(iii)   **(iv)**

2 (i) $(b \wedge p) \vee [b \wedge (\sim p \vee q)]$

(ii) $(a \wedge b) \vee (a \wedge c) \vee (b \wedge c) \vee (a \wedge b \wedge c)$

3 (i)   — $\sim 2^0$ —   (only one switch needed)

(ii)   — 2^0 —

(iii)   **(iv)**

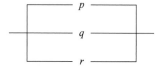 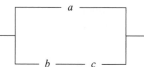

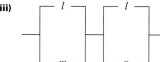 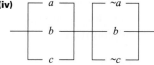

4

 or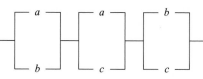

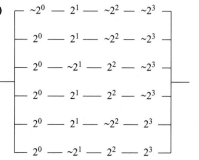

5

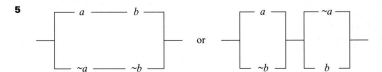

6

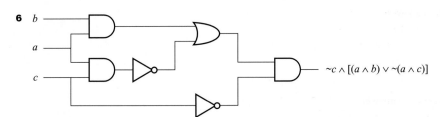

$\sim c \wedge [(a \wedge b) \vee \sim(a \wedge c)]$

7 $[p \wedge (\sim q \vee r)] \vee q$

8

switch
front door
back door

9

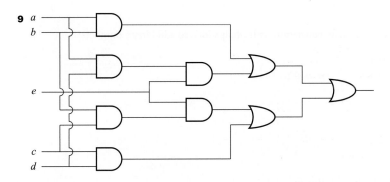

a
b

e

c
d

Exercise 4E (Page 120)

1 (i)

a	b	$a \Rightarrow b$	$a \wedge (a \Rightarrow b)$	$[a \wedge (a \Rightarrow b)] \Rightarrow a$
0	0	1	0	1
0	1	1	0	1
1	0	0	0	1
1	1	1	1	1

(ii)

p	q	$\sim p$	$\sim q$	$\sim p \vee \sim q$	$p \wedge \sim q$	$(\sim p \vee \sim q) \Rightarrow (p \wedge \sim q)$
0	0	1	1	1	0	0
0	1	1	0	1	0	0
1	0	0	1	1	1	1
1	1	0	0	0	0	1

(iii)

a

b

2 (i) (a) $(a \wedge \sim b) \vee (\sim a \wedge b)$

   (b) $(a \wedge \sim b) \vee (\sim a \wedge b) = [(a \wedge \sim b) \vee \sim a] \wedge [(a \wedge \sim b) \vee b]$
   $= (a \vee \sim a) \wedge (\sim b \vee \sim a) \wedge (a \vee b) \wedge (\sim b \vee b)$
   $= (\sim b \vee \sim a) \wedge (a \vee b) = (a \vee b) \wedge (\sim a \vee \sim b)$

(c)

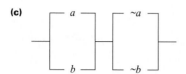

(ii) (a)

a	b	carry	answer
0	0	0	0
0	1	0	1
1	0	0	1
1	1	1	0

(b)

a	b	c	carry	answer
0	0	0	0	0
0	0	1	0	1
0	1	0	0	1
0	1	1	1	0
1	0	0	0	1
1	0	1	1	0
1	1	0	1	0
1	1	1	1	1

The circuit is a full adder and is used, in association with a half adder, to add binary numbers.

3 (i) a true $\Rightarrow b$ true but b true $\nRightarrow a$ true.

(ii)

n	s	$n \Rightarrow s$	$\sim s$	$\sim n$	$\sim s \Rightarrow \sim n$	$(n \Rightarrow s) \Leftrightarrow (\sim s \Rightarrow \sim n)$
0	0	1	1	1	1	1
0	1	1	0	1	1	1
1	0	0	1	0	0	1
1	1	1	0	0	1	1

(iii) $a \vee (\sim a \wedge b) = (a \vee \sim a) \wedge (a \vee b) = (a \vee b)$

4 (i) (a) Flicking a switch turns the light on if it is off, and off if it is on.

(b)

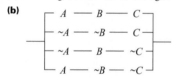

(ii)

5 (i) ──── C ────

(ii)

[A	$\wedge$	(C	$\Rightarrow$	A)]	$\Rightarrow$	C
0	0	0	1	0	1	0
0	0	1	0	0	1	1
1	1	0	1	1	0	0
1	1	1	1	1	1	1

Argument 1 is not valid.

[{(~	S	∨	L)	⇒	~	M}	∧	M}	⇒	S
1	0	1	0	1	1	0	0	0	1	0
1	0	1	0	0	0	1	0	1	1	0
1	0	1	1	1	1	0	0	0	1	0
1	0	1	1	0	0	1	0	1	1	0
0	1	0	0	1	1	0	0	0	1	1
0	1	0	0	1	0	1	1	1	1	1
0	1	1	1	1	1	0	0	0	1	1
0	1	1	1	0	0	1	0	1	1	1

Argument 2 is valid.

6 (i) (a)

(p	⇒	q)	⇔	~	(p	∧	~	q)
0	1	0	1	1	0	0	1	0
0	1	1	1	1	0	0	0	1
1	0	0	1	0	1	1	1	0
1	1	1	1	1	1	0	0	1

(b)

(~	q	⇒	~	p)	⇔	(p	⇒	q)
1	0	1	1	0	1	0	1	0
0	1	1	1	0	1	0	1	1
1	0	0	0	1	1	1	0	0
0	1	1	0	1	1	1	1	1

(ii)

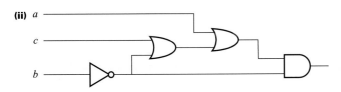

(iii) $[a \vee (\sim b \vee c)] \wedge \sim b = (a \wedge \sim b) \vee [(\sim b \vee c) \wedge \sim b]$
$= (a \wedge \sim b) \vee \sim b = \sim b$

7 (i)

a	∨	(b	∧	c)	⇔	(a	∨	b)	∧	(a	∨	c)
0	0	0	0	0	1	0	0	0	0	0	0	0
0	0	0	0	1	1	0	0	0	0	0	1	1
0	0	1	0	0	1	0	1	1	0	0	0	0
0	1	1	1	1	1	0	1	1	1	0	1	1
1	1	0	0	0	1	1	1	0	1	1	1	0
1	1	0	0	1	1	1	1	0	1	1	1	1
1	1	1	0	0	1	1	1	1	1	1	1	0
1	1	1	1	1	1	1	1	1	1	1	1	1

(ii) $\sim (p \wedge \sim q)$

(iii) $\sim [(v \wedge w) \vee (x \wedge y)]$

8 (i) $[(a \wedge b) \wedge (c \vee b)] \vee (a \vee b)$

(ii) $(a \wedge b) \wedge (c \vee b) = (a \wedge b \wedge c) \vee (a \wedge b \wedge b)$
$= (a \wedge b \wedge c) \vee (a \wedge b) = (a \wedge b)$
So $[(a \wedge b) \wedge (c \vee b)] \vee (a \vee b) = (a \wedge b) \vee (a \vee b) = (a \vee b)$

(iii)

[(a	∧	b)	∧	(c	∨	b)]	∨	(a	∨	b)	⇔	(a	∨	b)
0	0	0	0	0	0	0	0	0	0	0	1	0	0	0
0	0	0	0	1	1	0	0	0	0	0	1	0	0	0
0	0	1	0	0	1	1	1	0	1	1	1	0	1	1
0	0	1	0	1	1	1	1	0	1	1	1	0	1	1
1	0	0	0	0	0	0	1	1	1	0	1	1	1	0
1	0	0	0	1	1	0	1	1	1	0	1	1	1	0
1	1	1	1	0	1	1	1	1	1	1	1	1	1	1
1	1	1	1	1	1	1	1	1	1	1	1	1	1	1

9 (i) $(\sim s \Rightarrow \sim n) \wedge n$

(ii)

(~	s	⇒	~	n)	∧	n	⇒	s
1	0	1	1	0	0	0	1	0
1	0	0	0	1	0	1	1	0
0	1	1	1	0	0	0	1	1
0	1	1	0	1	1	1	1	1

(iii) $s \Rightarrow r$ so $\sim r \Rightarrow \sim s$

But $\sim s \Rightarrow \sim n$, so you can deduce that there is no snow and no north wind.

10 (i)

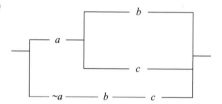

(ii) You could show this by an argument about majority voting, or by truth tables, or by Boolean algebra.

(iii)

```
   ┌─ a ─── b ──────┐
   │                │
───┤─── b ── c ─────├───
   │                │
   └─ a ─────── c ──┘
```

(iv) ─── a ─── b ─── c ───

11 (i) (a)

a	b	carry	answer
0	0	0	0
0	1	0	1
1	0	0	1
1	1	1	0

(b)

a	b	c	carry	answer
0	0	0	0	0
0	0	1	0	1
0	1	0	0	1
0	1	1	1	0
1	0	0	0	1
1	0	1	1	0
1	1	0	1	0
1	1	1	1	1

(ii) The 'A' and the '2'

Chapter 5

❷ (Page 128)

Only one arc can be undirected, and that is indicated in the diagram. All other arcs represent pipes which are either taking water away from S, or to T.

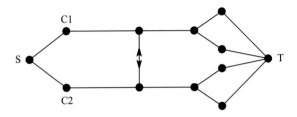

❷ (Page 129)

The flow pattern is feasible.

SA: capacity = 13; flow = 13; OK
SB: capacity = 12; flow = 8; OK
AC: capacity = 10; flow = 10; OK
AD: capacity = 5; flow = 3; OK
BC: capacity = 5; flow = 3; OK
BD: capacity = 5; flow = 5; OK
CT: capacity = 13; flow = 13; OK
DT: capacity = 12; flow = 8; OK

Vertex A: in = 13; out = 10 + 3 = 13; OK
Vertex B: in = 8; out = 3 + 5 = 8; OK
Vertex C: in = 3 + 10 = 13; out = 13; OK
Vertex D: in = 3 + 5 = 8; out = 8; OK

❷ (Page 130)

See page 131.

❷ (Page 131)

The flow pattern is feasible.

❷ (Page 132)

See text that follows.

Exercise 5A (Page 136)

1 (i)   Source = B, Sink = E
  (ii) 27
  (iii) e.g. BA = 8; BC = 7; BE = 12; AD = 8; CD = 2; CF = 5; DH = 10;
        FH = 0; FG = 2; FE = 3; HG = 10; GE = 12
  (iv) Total flow (e.g. BA + BC + BE) = 27. This is the same as the capacity of the cut in part (ii), so it must be a maximal flow.

2 (i) 20

(ii) e.g. SA = 10; SC = 10; AB = 10; CA = 0; CB = 5; CT = 5; BT = 15
Established flow = 20 = capacity of a cut, so flow is maximal.

(ii) Cut now has capacity 30.
Can now achieve a total flow of 22.
e.g. SA = 12; SC = 10; AB = 10; AC = 2; CB = 5; CT = 7; BT = 15
This can be seen to be maximal by virtue of cut S | ABCT.

3 (i) Needs SS$_1$ with capacity 20 and SS$_2$ with capacity 30.

(ii) Needs T$_1$T with capacity 25 (or more), T$_2$T with capacity 25 (or more) and T$_3$T with capacity 20 (or more).

(iii) Cut SS$_1$S$_2$A | BCDT$_1$T$_2$T$_3$T has capacity 47. There are many ways of achieving a flow of 47.

4 (i) (a) SABCE | DT, capacity = 15

(b) SABC | DET, capacity = 14

(ii) (a) SA = 2; SB = 6; SC = 7; AD = 2; AB = 0; BD = 5; BE = 1; CB = 0; CE = 7; DT = 7; ET = 8

(b) As above, but with SB = 5; EB = 0; ET = 7

5 (i)

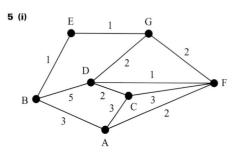

(ii) BE = 0; BD = 5; BA = 3; EG = 0; DG = 2; DF = 1; DC = 2; AC = 1; AF = 2; GF = 2; CF = 3
Cut BEDAGC | F has capacity 8.

❼ (Page 139)

There are $2^9 = 512$ subgraphs altogether, since there are 9 edges in the bipartite graph and each may either be in, or not in, the subgraph. But the majority of these subgraphs are not matchings.
There is 1 matching containing 0 edges.
There are 9 matchings containing 1 edge.
There are 23 matchings containing 2 edges.
There are 21 matchings containing 3 edges.
There are 2 matchings containing 4 edges.

Exercise 5B (Page 141)

1 e.g. First alternating path = E–400–D–1500.
So replace D–400 by E–400 and D–1500.
Second alternating path = F–100–A–200
So replace A–100 by F–100 and A–200
Gives A–200; B–800; C–hurdles; D–1500; E–400; F–100

2 Only one: A–Wed; B–Mon; C–Tues; D–Thurs; E–Fri

3 (i) e.g. 7–C–4–F, giving A–1; B–3; C–7; E–5; F–4; G–2; H–6

(ii) D–1–A–3–B–6, giving A–3; B–6; C–7; D–1; E–5; F–4; G–2

4 (i) e.g. 1–C–2–A, giving A–2; D–3; C–1

(ii) e.g. SA = 1; SB = 0; SC = 1; SD = 1; A2 = 1; B2 = 0; C1 = 1; C2 = 0; C3 = 0; D3 = 1; 1T = 1; 2T = 1; 3T = 1
Flow pattern will imply a solution to the matching problem if flows are not fractional.

5 A–1; B–4; C–2; D–3
   A–1; B–4; C–2; E–3
   A–4; D–1; B–2; C–3
   A–4; D–1; B–2; E–3
   A–4; D–1; C–2; E–3
   B–4; D–1; C–2; E–3
   B–4; D–1; A–2; C–3
   B–4; D–1; A–2; E–3

Exercise 5C (Page 144)

1 (i) Maximum total flow = 19
    e.g. SA = 10; SB = 9; AC = 14; DA = 4; BD = 9; CT = 14; DT = 5

  (ii) S | ABCDT; capacity of cut = 19 so flow is maximal

2 (i) SABD | CT, capacity = 22

  (ii) Flow = 19 < 22

  (iii) S → A → D → B → C → T (augmentation of 3 possible)
      SA = 10; AD = 5; BD = 4; BC = 8; CT = 13; others unchanged.

3 (i) Source = B, sink = F

  (ii) A partition of the vertices into two sets, one containing the source, S, and the other containing the sink, T.
    Capacity of given cut = 43.

  (iii) B → C → H → G → F with an augmentation of 4

  (iv) No, B → C → H → G → E → F gives a further augmentation of 1.
    BA = 25; BC = 18; AH = 12; AE = 13; HC = 0; CD = 18; HG = 12; DG = 8; DE = 10; GE = 5; GF = 15; EF = 28

  (v) Capacity of given cut = 43 = established flow

4 (iii) Maximal flow = 27
    e.g. SA = 8; SB = 12; SC = 7; AP = 8; BP = 6; BQ = 6; CQ = 7; PX = 14; QX = 4; QY = 4; QR = 5; RY = 5;
    XT = 18; YT = 9

  (iv) SBCQR | APXYT

  (v) 8 supplied from A, 12 from B and 7 from C
    18 supplied to X and 9 to Y

  (vi) Augmentation of 3 available to give a new maximum of 30.
    Labelling of QP after augmentation:

5		5̶	2
0̶	3	5̶	8

  (vii) Introduce an arc with large capacity from R to the sink T.

5 (i)

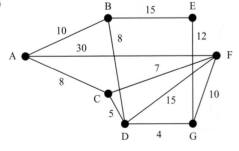

  (ii) BAF = 2; BDCF = 2; BDF = 2; BDGF = 1; BEGF = 1; BACF = 1; total flow = 9.
    Cut BDGE | AFC has capacity 9.

  (iii) BDCF; length = 20

  (iv) $22\frac{2}{9}\%$

6 (i) SA = 9; SC = 18; AB = 14; CA = 5; CB = 5; BT = 19; CT = 8.

(ii) SC | ABT; capacity = 27; proves that established flow is maximal.

(iii) 20, with SC = 11

7 (i) SBD | ACT; capacity = 24

Established flow = 18 < 24

Maximum flow: SA = 12; SB = 12; AC = 12; AD = 0; BD = 12; CT = 12; DT = 12.

(ii) Previous minimum cut now has capacity 30.

New minimum cut = SB | ACDT, with capacity 26.

Maximum flow: SA = 12; SB = 14; AC = 14; DA = 2; BD = 14; CT = 14; DT = 12.

8 (i)

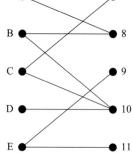

(ii) Initial matching

e.g.

 8 breakthrough, giving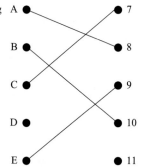

thence D —— 10 —— B —— 8 —— A —— 7 —— C end – no breakthrough

A complete matching is not possible because D would have to be watched at 10 pm, leaving A, B and C to be watched at 7 pm or 8 pm.

Chapter 6

⌨ **(Page 153)**

2	MAX	SA + SB + SC + SD		OBJECTIVE FUNCTION VALUE	
	ST	$SA - A1 - A4 = 0$		1)	4.000000
		$SB - B1 - B3 = 0$		VARIABLE	VALUE
		$SC - C2 - C3 = 0$		SA	1.000000
		$SD - D3 = 0$		SB	1.000000
		$A1 + B1 - T1 = 0$		SC	1.000000
		$C2 - T2 = 0$		SD	1.000000
		$B3 + C3 + D3 - T3 = 0$		A1	0.000000
		$A4 - T4 = 0$		B3	0.000000
		$SA <= 1$		T1	1.000000
		$SB <= 1$		T2	1.000000
		$SC <= 1$		T3	1.000000
		$SD <= 1$		T4	1.000000
		$T1 <= 1$		A4	1.000000
		$T2 <= 1$		B1	1.000000
		$T3 <= 1$		C2	1.000000
		$T4 <= 1$		C3	0.000000
	END			D3	1.000000

❷ **(Page 153)**

There doesn't appear to be anything to prevent the possibility of fractional answers but the simplex algorithm always produces 0/1 solutions.

❷ **(Page 154)**

The first four constraints say that each machine must be employed. The next four constraints say that each task must be completed.

Exercise 6A (Page 160)

1 MIN $15AB + 8AC + 8CA + 5AD + 5DA + 21EA + 2AF + 2FA + 11CB + 23DB + 40EB$
$+ 17FB + 9CD + 9DC + 26EC + 12CF + 12FC + 17ED + 11DF + 11FD + 13EF$

ST $EA + EB + EC + ED + EF = 1$
$CA + DA + EA + FA - AB - AC - AD - AF = 0$
$AD + CD + ED + FD - DA - DB - DC - DF = 0$
$AF + CF + DF + EF - FA - FB - FC - FD = 0$
$AB + CB + DB + EB + FB = 1$

END
Gives $E \rightarrow F \rightarrow A \rightarrow B$ with length 30.

2 MAX $SB + SC$

ST $SB + CB + DB + EB - BC - BD - BE = 0$
$SC + BC + DC + EC - CB - CD - CE = 0$
$BD + CD + ED - DB - DC - DE - DT = 0$
$BE + CE + DE - EB - EC - ED - ET = 0$
$SB <= 15$
$SC <= 8$
$BC <= 11$
$CB <= 11$

$$BD \leq 23$$
$$DB \leq 23$$
$$BE \leq 4$$
$$EB \leq 4$$
$$CD \leq 9$$
$$DC \leq 9$$
$$CE \leq 6$$
$$EC \leq 6$$
$$DE \leq 17$$
$$ED \leq 17$$
$$DT \leq 11$$
$$ET \leq 13$$

END

This achieves a maximum flow of 23. There are many alternative solutions and which is delivered depends on the initial state of the software. It may deliver an answer in which, for instance, $DE = 17$ and $ED = 8$. This is equivalent to a flow of 9 from D to E.

3 MIN $\quad 10A1 + 15A2 + 8A3 + 5A4 + 21A5 + 2A6 + 15B1 + 5B2 + 11B3 + 23B4 + 4B5 + 17B6 + 8C1 + 11C2 + 9C4$
$+ 6C5 + 12C6 + 5D1 + 23D2 + 9D3 + 30D4 + 17D5 + 11D6 + 21E1 + 4E2 + 6E3 + 17E4 + 11E5 + 13E6$
$+ 2F1 + 17F2 + 12F3 + 11F4 + 13F5 + 7F6$

ST $\quad A1 + A2 + A3 + A4 + A5 + A6 = 1$
$\quad\quad B1 + B2 + B3 + B4 + B5 + B6 = 1$
$\quad\quad C1 + C2 + C3 + C4 + C5 + C6 = 1$
$\quad\quad D1 + D2 + D3 + D4 + D5 + D6 = 1$
$\quad\quad E1 + E2 + E3 + E4 + E5 + E6 = 1$
$\quad\quad F1 + F2 + F3 + F4 + F5 + F6 = 1$
$\quad\quad A1 + B1 + C1 + D1 + E1 + F1 = 1$
$\quad\quad A2 + B2 + C2 + D2 + E2 + F2 = 1$
$\quad\quad A3 + B3 + C3 + D3 + E3 + F3 = 1$
$\quad\quad A4 + B4 + C4 + D4 + E4 + F4 = 1$
$\quad\quad A5 + B5 + C5 + D5 + E5 + F5 = 1$
$\quad\quad A6 + B6 + C6 + D6 + E6 + F6 = 1$

END

Best allocation: A → 4, B → 5, C → 3, D → 1, E → 2, F → 6, at a cost of 25.

4 100 items available and 88 required so a dummy demand of 12 is needed. This will show where items are left.

Formulation as for question 3, but with the following constraints.
$A1 + A2 + A3 + A4 + A5 + A6 + ADum = 15$
$B1 + B2 + B3 + B4 + B5 + B6 + BDum = 23$
$C1 + C2 + C3 + C4 + C5 + C6 + CDum = 14$
$D1 + D2 + D3 + D4 + D5 + D6 + DDum = 12$
$E1 + E2 + E3 + E4 + E5 + E6 + EDum = 24$
$F1 + F2 + F3 + F4 + F5 + F6 + FDum = 12$
$A1 + B1 + C1 + D1 + E1 + F1 = 6$
$A2 + B2 + C2 + D2 + E2 + F2 = 32$
$A3 + B3 + C3 + D3 + E3 + F3 = 12$
$A4 + B4 + C4 + D4 + E4 + F4 = 16$
$A5 + B5 + C5 + D5 + E5 + F5 = 10$
$A6 + B6 + C6 + D6 + E6 + F6 = 12$
$ADum + BDum + CDum + DDum + EDum + FDum = 12$

Gives $A4 = 14$, $A6 = 1$, $B2 = 8$, $B5 = 10$, $C3 = 12$, $C4 = 2$, $D1 = 5$, $E2 = 24$, $F1 = 1$, $F6 = 11$ at a cost of 370 with 5 left at B and 7 left at D.

5 MAX   M

ST $M - 10P1 - 15P2 - 8P3 - 5P4 - 21P5 - 2P6 <= 0$

        $M + 15P1 - 5P2 - 11P3 - 23P4 - 4P5 - 17P6 <= 0$

        $M + 16P1 + 11P2 - 9P4 - 6P5 - 12P6 <= 0$

        $M + 8P1 + 3P2 - 9P3 - 30P4 - 17P5 - 11P6 <= 0$

        $M + 12P1 + 14P2 - 6P3 + 17P4 + 11P5 - 13P6 <= 0$

        $M - 2P1 - 17P2 + 20P3 + 9P4 + 23P5 - 7P6 <= 0$

        $P1 <= 1$

        $P2 <= 1$

        $P3 <= 1$

        $P4 <= 1$

        $P5 <= 1$

        $P6 <= 1$

        $P1 + P2 + P3 + P4 + P5 + P6 = 1$

END

FREE M

Gives $M = 6.033315$ by taking $P2 = 0.176952$, $P5 = 0.091207$ and $P6 = 0.731841$.

6 MIN   T

ST $TC - TA >= 15$

        $TC - TB >= 10$

        $TD - TA >= 15$

        $TE - TA >= 15$

        $TE - TB >= 10$

        $TF - TD >= 2$

        $TF - TE >= 8$

        $TG - TA >= 15$

        $TH - TC >= 5$

        $TH - TG >= 3$

        $T - TF >= 12$

        $T - TH >= 13$

END

This gives a duration (T) of 35.

❼ (Page 162)

Suppose that x is to lie between 0 and 100, and that the threshold value is 0.1.

Then the inequalities are $IX \leq 10x$ and $IX \geq 0.01x - 0.001$.

Thus if, for instance, $x = 0.05$, then $IX \leq 0.5$ and $IX \geq -0.0005$, forcing IX to be 0.

If, for instance, $x = 0.2$, then $IX \leq 2$ and $IX \geq 0.001$, forcing IX to be 1.

If, for instance, $x = 99$, then $IX \leq 990$ and $IX \geq 0.999$, again forcing IX to be 1.

Exercise 6B (Page 164)

1 MAX   SA + SB   MAX   SA + SB

ST SA + DA − AD − AC = 0	ST SA − AC − AD = 0
AC − CT = 0	AC + BC − CT = 0
SB − BD = 0	SB − BC − BD = 0
AD + BD − DA − DT = 0	AD + BD − DT = 0
SA <= 10	SA <= 12
SB <= 9	SB <= 12
AC <= 16	AC <= 5
AD <= 6	AD <= 9
DA <= 6	BC <= 8
BD <= 10	BD <= 7
CT <= 17	CT <= 15
DT <= 5	DT <= 9
END	END

2 MAX   BA + BD + BE

ST   BA + CA − AC − AF = 0

AC + DC − CA − CD − CF = 0

BD + CD + GD − DC − DF − DG = 0

BE + GE − EG = 0

DG + EG − GD − GF = 0

BA <= 3

AC <= 3

CA <= 3

AF <= 2

BD <= 6

DB <= 6

CD <= 2

DC <= 2

CF <= 4

DF <= 2

DG <= 1

GD <= 1

EG <= 2

FG <= 2

GF <= 2

END

MIN   10BA + 8BD + 15BE + 8AC + 8CA + 30AF + 30FA + 5CD + 5DC + 7CF + 15DF + 4DG + 4GD
+ 12EG + 12GE +10GF

ST   BA + BD + BE = 1

BA + CA − AC − AF = 0

AC + DC − CA − CD − CF = 0

BD + CD + GD − DC − DF − DG = 0

BE + GE − EG = 0

G + EG − GD − GF = 0

AF + CF + DF + GF = 1

END

3 MAX   A1 + A2 + B8 + B15 + C2 + CH + D4 + D8 + D15 + E4 + E8 + F1 + FH

    ST   A1 + A2 <= 1

           B8 + B15 <= 1

           C2 + CH <= 1

           D4 + D8 + D15 <= 1

           E4 + E8 <= 1

           F1 + FH <= 1

           A1 + F1 <= 1

           A2 + C2 <= 1

           D4 + E4 <= 1

           B8 + D8 + E8 <= 1

           B15 + D15 <= 1

           CH + FH <= 1

   END

   MAX   A2 + B2 + C1 + C2 + C3 + D3

    ST   A2 <= 1

           B2 <= 1

           C1 + C2 + C3 <= 1

           D3 <= 1

           C1 <= 1

           A2 + B2 + C2 <= 1

           C3 + D3 <= 1

   END

4 Allocation: A → 1, B → 4, C → 3 and D → 2, with a time of 238.

Transportation: 5 from W_1 to S_2, 9 from W_2 to S_1, 3 from W_2 to S_4, 4 from W_3 to S_2, 9 from W_3 to S_3 and 4 from W_3 to S_4, at a cost of 102.

5 Make 5 trips using a 15-tonne lorry and 25 trips using a 20-tonne lorry.
Total cost = £1385.

Exercise 6C (Page 168)

1 Make 4 tonnes of X and 2 tonnes of Y at a profit of £880. This uses all of the finance and all of the staff time, but leaves 5 m³ of storage space unused.

The marginal value of finance is 40p per £.

The marginal value of staff time is £10 per hour.

The profit per tonne of X could be anything between £120 and £240 without changing the optimal basis (i.e. with it still being best to use all of the finance and all of the staff time).

Similarly the profit per tonne of Y could be anything between £80 and £160, the available finance between £750 and £1200, the available staff time between 40 and 56 hours, and the storage space anything above 10 m³.

2 Use 4 kg of X and 1 kg of Y at a cost of £5. This solution just satisfies constraint numbers 2 and 3, and leaves a slack of 1 in constraint number 1.

The marginal cost of constraint number 2 is 25p per unit, as is the marginal cost of constraint number 3.

This solution (i.e. having constraints 2 and 3 tight) holds when X costs between 50p and £1.50 per kg, and when Y costs between 67p and £2 per kg. It holds for constraint right-hand sides respectively less than 9, between 12.67 and 18 and between 4.67 and 10. (These are all the limits of changes in just one parameter whilst the others are held constant.)

3 Make 450 chairs and 10 tables, giving an income of £39 500.

The marginal cost of the budget constraint is £2 per £.

The marginal cost of the time constraint is £10 per hour.

This solution, using all of the budget and all of the time, is valid for chair prices between £70 and £93.33 and for table prices between £300 and £400. Similarly the basis is unchanged for budgets between £9750 and £13 000 and for time between 1500 and 2000 hours.

Exercise 6D (Page 168)

1 C runs leg 1, A runs leg 2, D runs 3 and B runs 4; 250 seconds.

2 (i) $A \rightarrow B \rightarrow C \rightarrow D$ with weight 8.

(ii) must leave A
if arrives at B then must leave
if arrives at C then must leave
must arrive at D

3 Allocation: $A \rightarrow 6$, $B \rightarrow 1$, $C \rightarrow 5$, $D \rightarrow 2$, $E \rightarrow 4$, $F \rightarrow 3$, at a total cost of £8300.

4 (i) $2 + 4 + 1 + 4 + 2 = 13$ hours

(ii) Total time from the depot, at the jobs and back to the depot, is fixed. So to minimise the total time it is only necessary to minimise the transfer time, a straightforward allocation problem.
However, it might be thought best to minimise the maximum time. This is a minimax problem. As in the allocation problem variables such as AY are used, and the usual allocation constraints apply. But the objective is to minimise M where the extra constraints $M \geqslant 16AY$, etc. are included. It needs integer variables.
To minimise total time pair A with Y, B with X, C with W and D with Z.
To minimise maximum time one solution is to pair A with W, B with Z, C with X and D with Y (but see below).

(iii) If minimising the total time then that time is 51 hours with the longest any one machine is away being 16 hours. If minimising the maximum time that a machine is away then this is 15 hours, but the total time then goes up to 54 hours (given the above solution).
However, after discovering that the minimax time is 15, you can return to minimising the total time with the constraint that no machine must be away for more than 15 hours. This gives A with W, B with X, C with Z and D with Y, with a total time of 52 hours.

5 MIN $10000T + 1000SA + 750SB + 1500SC + 1000SE + 250SF + 500SG + 1500SH$
ST $TC - TA + SA >= 15$
        $TC - TB + SB >= 10$
        $TD - TA + SA >= 15$
        $TE - TA + SA >= 15$
        $TE - TB + SB >= 10$
        $TF - TD >= 2$
        $TF - TE + SE >= 8$
        $TG - TA + SA >= 15$
        $TH - TC + SC >= 5$
        $TH - TG + SG >= 3$
        $T - TF + SF >= 12$
        $T - TH + SH >= 13$
        $SA <= 2$
        $SB <= 3$
        $SC <= 2$
        $SE <= 1$
        $SF <= 3$
        $SG <= 1$
        $SH <= 1$
END

Duration = 29 days. Extra cost = £6750 by shortening A by 2 days, C by 2 days, E by 1 day and F by 3 days.

6 (i) 20 000 000ZA is the cost of building factory A, or not.

3.5AX is the cost of manufacturing AX components at A and shipping them to X. Similarly, 4AY is the cost of manufacturing AY components at A and shipping them to Y.

Minimise $\quad$ 20 000 000ZA + 40 000 000ZB + 30 000 000ZC

$$+ 3.5AX + 4AY + 2.4BX + 3.2BY + 2.7CX + 3.3CY$$

(ii) AX + AY − kZA ⩽ 0 ensures that if anything is produced at factory A then the factory has to be built. k has to be at least the sum of the maximum possible values which AX and AY can take.

AX + AY − 15 000 000ZA ⩽ 0

BX + BY − 25 000 000ZB ⩽ 0

CX + CY − 20 000 000ZC ⩽ 0

(iii) AX + BX + CX ⩾ 30 000 000

AY + BY + CY ⩾ 10 000 000

(iv) Build factories B and C. Supply 25 million components from B to X, 5 million from C to X and 10 million from C to Y, at a total cost of £176.5 million

7 (i) The line segment joining (x_1, y_1) to (x_2, y_2).

(ii) The inside and boundary of the triangle with the three points as vertices.

(iii) The lower boundary of the region formed by triangles defined by triples of points.

8 (i) 1 and 2.6

$ap + bq + c$ will be > 0 for points on one side of the line and < 0 for points on the other side. The modulus function, together with the division by $\sqrt{a^2 + b^2}$, gives the distance.

(ii) $0.6x + 0.8y = 4$, $x = 0$ and $y = 0$.

(iii) The required point is $(1\frac{2}{3}, 1\frac{2}{3})$. The minimax distance of the point from each line is $1\frac{2}{3}$, and in this problem that means that each of the three distances is $1\frac{2}{3}$.

(iv) If the point were further from one line than from another, then the minimax distance could be reduced by moving the point towards the more distant line and away from the other. This will always be possible provided that there is no parallel pair.

(v) In the general case, with many lines, it will not be clear on which side of each line the solution lies. Thus, in the example, if the point is on the origin side of $0.6x + 0.8y = 4$ then the second inequality is redundant. But if the point is on the other side then the first inequality will be inactive and the second will impose the required condition that $z \geqslant -4 + 0.6x + 0.8y$.

9 In the following formulation F1 is an indicator variable, indicating whether or not a fire station is built in city 1. X23, for instance, is also an indicator variable, indicating whether or not city 2 is to be served by a fire station in city 3. C1 is the time taken for a fire engine to reach city 1.

```
MIN    F1 + F2 + F3 + F4 + F5 + F6
ST     6F1 − X11 − X21 − X31 − X41− X51 − X61 >= 0
       6F2 − X12 − X22 − X32 − X42 − X52 − X62 >= 0
       6F3 − X13 − X23 − X33 − X43 − X53 − X63 >= 0
       6F4 − X14 − X24 − X34 − X44 − X54 − X64 >= 0
       6F5 − X15 − X25 − X35 − X45 − X55 − X65 >= 0
       6F6 − X16 − X26 − X36 − X46 − X56 − X66 >= 0
       C1 − 10X12 − 20X13 − 30X14 − 30X15 − 20X16 = 0
       C2 − 10X21 − 25X23 − 35X24 − 20X25 − 10X26 = 0
       C3 − 20X31 − 25X32 − 15X34 − 30X35 − 20X36 = 0
       C4 − 30X41 − 35X42 − 15X43 − 15X45 − 25X46 = 0
       C5 − 30X51 − 20X52 − 30X53 − 15X54 − 14X56 = 0
       C6 − 20X61 − 10X62 − 20X63 − 25X64 − 14X65 = 0
       X11 + X12 + X13 + X14 + X15 + X16 = 1
       X21 + X22 + X23 + X24 + X25 + X26 = 1
       X31 + X32 + X33 + X34 + X35 + X36 = 1
```

$$X41 + X42 + X43 + X44 + X45 + X46 = 1$$
$$X51 + X52 + X53 + X54 + X55 + X56 = 1$$
$$X61 + X62 + X63 + X64 + X65 + X66 = 1$$
$$C1 <= 15$$
$$C2 <= 15$$
$$C3 <= 15$$
$$C4 <= 15$$
$$C5 <= 15$$
$$C6 <= 15$$
END
INT   42

A solution is 2, with $F2 = 1$ and $F4 = 1$.

Alternatively, you can replace the table of times by a table showing which times are less than or equal to 15 as shown.

	City 1	City 2	City 3	City 4	City 5	City 6
City 1	1	1	0	0	0	0
City 2	1	1	0	0	0	0
City 3	0	0	1	1	0	0
City 4	0	0	1	1	1	0
City 5	0	0	0	1	1	1
City 6	0	1	0	0	1	1

The following formulation, given by looking down columns, or across rows, then solves the problem.

MIN   C1 + C2 + C3 + C4 + C5 + C6
ST   C1 + C2 >= 1
       C1 + C2 + C6 >= 1
       C3 + C4 >= 1
       C3 + C4 + C5 >= 1
       C4 + C5 + C6 >= 1
       C2 + C5 + C6 >= 1
END

10 (i) Letting x, y and z be the numbers of large, medium and small lorries respectively:

Minimise $8000x + 4000y + 3200z$
Subject to $4000x + 2400y + 2000z \geqslant 200\,000$
$$x + y + z \leqslant 65$$
$$z \geqslant 20$$

gin 3

(ii) OBJECTIVE FUNCTION VALUE

1   374400.000

VARIABLE	VALUE
X	33.000000
Y	10.000000
Z	22.000000
ROW	SLACK OR SURPLUS
2)	0.000000
3)	0.000000
4)	2.000000

The company should hire 33 large, 10 medium and 22 small lorries, at a cost of £374 400, using all the drivers.

(iii) Deleting this requirement gives a solution of (28, 35, 2) at a cost of £370 400. So the cost of the requirement is £4000.

(iv) Letting x_h and x_o be the numbers of hired and owned large lorries respectively, etc. and d be the indicator variable:

Minimise $\quad 8000x_h + 4000y_h + 3200z_h + 6000x_o + 3000y_o + 2400z_o + 50\,000d$

Subject to $\quad 4000x_h + 2400y_h + 2000z_h + 4000x_o + 2400y_o + 2000z_o >= 200\,000$

$$x_h + y_h + z_h + x_o + y_o + z_o <= 65$$
$$z_h + z_o >= 20$$
$$(x_o + y_o + z_o <= 50) - \text{this constraint is not needed}$$
$$50d - x_o - y_o - z_o >= 0$$

end

gin 6

int d

(v) OBJECTIVE FUNCTION VALUE

1)   342800.0

VARIABLE	VALUE
XH	0.000000
YH	0.000000
ZH	15.000000
XO	33.000000
YO	10.000000
ZO	7.000000
D	1.000000

ROW	SLACK OR SURPLUS
2)	0.000000
3)	0.000000
4)	2.000000
5)	0.000000
6)	50.000000

The company should hire 15 small lorries and use 33 large, 10 medium and 7 small lorries of its own. It is worth setting up the depot, the saving being £374 400 – £342 800 = £31 600.

11 (i) The xs represent the proportion of the maximum power output being used from each of stations 1, 2 and 3. The hourly cost of running station 1 is $x_1 \times 4 \times 0.75$, etc. and the objective is to minimise the sum of these three hourly costs. The second constraint sets the supply equal to the demand.

(ii) OBJECTIVE FUNCTION VALUE

1)   4.595

VARIABLE	VALUE	REDUCED COST
X1	0.000000	0.080000
X2	1.000000	0.000000
X3	0.214286	0.000000
DEMAND	6.500000	0.000000

The island should run station 2 at maximum output and station 3 at $\frac{3}{14}$ of maximum, at a cost of 4.595 monetary units per hour.

(iii) OBJECTIVE FUNCTION VALUE

1)   3.290000

VARIABLE	VALUE	REDUCED COST
X1	0.000000	0.200000
X2	0.940000	0.000000
X3	0.000000	0.200000
DEMAND	4.700000	0.000000

The island should run station 2 only, at 94% of capacity, at a cost of 3.29 monetary units per hour.

(iv) Minimise $3X1 + 3.5X2 + 5.11X3 + 4.32X4 + 3.04X5 + 2.07X6$
$+ 6.16X7 + 3.33X8 + 4.712X9 + 6.57X10$

Subject to demand $= 38.2$

$4X1 + 5X2 + 7X3 + 6X4 + 4X5 + 3X6 + 8X7$
$+ 4.5X8 + 6.2X9 + 9X10 - \text{demand} = 0$

$X1 >= 0$

$X1 <= 1$

etc.

OBJECTIVE FUNCTION VALUE

1) 27.67500

VARIABLE	VALUE	REDUCED COST
X1	0.925000	0.000000
X2	1.000000	0.000000
X3	1.000000	0.000000
X4	1.000000	0.000000
X5	0.000000	0.040000
X6	1.000000	0.000000
X7	0.000000	0.160000
X8	1.000000	0.000000
X9	0.000000	0.062000
X10	1.000000	0.000000
DEMAND	38.200001	0.000000

(v) The objective changes to:

$\ldots + 2.4X11 + 4.7I$

The constraints become:

$\ldots + 10X11$

$X11 >= 0$

$X11 <= 1$

$I - X11 >= 0$

int I

OBJECTIVE FUNCTION VALUE

1)   27.35600

VARIABLE	VALUE	REDUCED COST
I	1.000000	−0.200000
X1	0.000000	0.080000
X2	1.000000	0.000000
X3	1.000000	0.000000
X4	1.000000	0.000000
X5	0.000000	0.120000
X6	1.000000	0.000000
X7	0.000000	0.320000
X8	0.000000	0.045000
X9	0.000000	0.186000
X10	0.800000	0.000000
X11	1.000000	0.000000
DEMAND	38.200001	0.000000

(vi) The minimum demand is 13 units. This can be found either by iteration or by setting

$3 \times 0.69 + 5 \times 0.7 + 5 \times 0.72 = 4.7 + 2.4 + 3 \times 0.69$

12 (i) (a) The variables represent the numbers of full-sized rolls to be cut according to each cutting plan.

The constraints ensure that there are at least 150 4 m rolls, 200 3 m rolls and 100 2 m rolls.

The LP gives $x_1 = 150$ and $x_4 = 50$. So 200 rolls are used, 150 of which are cut according to plan 1 and 50 according to plan 4.

(b) As above. The result 'objective $= 0$' means that there is no trim loss.

(ii) (a)

Cutting plan	4 m rolls	3 m rolls	2 m rolls	1.6 m rolls	Trim loss (m)
1	1	1	0	0	0
2	1	0	1	0	1
3	1	0	0	1	1.4
4	0	2	0	0	1
5	0	1	2	0	0
6	0	1	1	1	0.4
7	0	1	0	2	0.8
8	0	0	3	0	1
9	0	0	2	1	1.4
10	0	0	1	3	0.2
11	0	0	0	4	0.6

(b) Minimise $\quad x_1 + x_2 + x_3 + x_4 + x_5 + x_6 + x_7 + x_8 + x_9 + x_{10} + x_{11}$

Subject to $\quad x_1 + x_2 + x_3 \geqslant 150$

$x_1 + 2x_4 + x_5 + x_6 + x_7 \geqslant 100$

$x_2 + 2x_5 + x_6 + 3x_8 + 2x_9 + x_{10} \geqslant 250$

$x_3 + x_6 + 2x_7 + x_9 + 3x_{10} + 4x_{11} \geqslant 300$

end

gin 11

This gives $x_1 = 100$, $x_2 = 50$, $x_8 = 34$ and $x_{10} = 100$, with a trim loss of 104 m (or other combinations with $x_1 + x_2 = 150$, with slightly different trim losses). You can minimise the trim loss with the objective:

Minimise $\quad x_2 + 1.4x_3 + x_4 + 0.4x_6 + 0.8x_7 + x_8 + 1.4x_9 + 0.2x_{10} + 0.6x_{11}$

This gives $x_1 = 150$, $x_5 = 75$ and $x_{10} = 100$, with a trim loss of 20 m.

13 (i) 3 000 000 is the fixed cost per annum of running each large plane. The variable costs are $8000 \times 2.9 = 23\,200$ for each long-haul flight, etc.

(ii) The right-hand side of the inequality presents the large plane days available, and the left-hand side the large plane days used.

(iii) MIN $\quad 3\,000\,000$NL $+ 23\,200$LL $+ 14\,500$TL $+ 2900$SL

$+ 3\,000\,000$NM $+ 16\,000$LM $+ 10\,000$TM $+ 2000$SM

$+ 1\,500\,000$NS $+ 16\,000$LS $+ 10\,000$TS $+ 2000$SS

ST $\quad$ LL $+ 0.5$TL $+ 0.25$SL $- 300$NL $\leqslant 0$

LM $+ 0.5$TM $+ 0.25$SM $- 300$NM $\leqslant 0$

LS $+ 0.5$TS $+ 0.25$SS $- 300$NS $\leqslant 0$

250LL $+ 200$LM $+ 150$LS $\geqslant 500\,000$

250TL $+ 200$TM $+ 150$TS $\geqslant 1\,000\,000$

250SL $+ 200$SM $+ 150$SS $\geqslant 2\,250\,000$

18NL $+ 15$NM $+ 12$NS $\leqslant 500$

END

GIN $\quad$ 12

(iv) OBJECTIVE FUNCTION VALUE

1) 0.1891340E+09

VARIABLE	VALUE	REDUCED COST
NL	5.000000	3000000.00
LL	0.000000	23200.0000
TL	0.000000	14500.0000
SL	6000.000	2900.00000
NM	17.00000	3000000.00
LM	2500.000	16000.0000
TM	5000.000	10000.0000
SM	399.0000	2000.00000
NS	4.000000	1500000.00
LS	0.000000	16000.0000
TS	0.000000	10000.0000
SS	4468.000	2000.00000

ROW	SLACK OR SURPLUS	DUAL PRICES
2)	0.000000	0.000000
3)	0.250000	0.000000
4)	83.00000	0.000000
5)	0.000000	0.000000
6)	0.000000	0.000000
7)	0.000000	0.000000
8)	107.0000	0.000000

NO. ITERATIONS = 267

Virginia should buy 5 large planes (for 6000 short-haul flights), 17 medium-sized planes (for 2500 long-haul, 5000 transatlantic and 399 short-haul flights) and 4 small planes (for 4468 short-haul flights), at an annual cost of $189m.

Capital cost = $393m

There will be 0.25 days slack on the medium-sized planes and 83 days slack on the small ones.

(v) Is every flight full to capacity? Scheduling issues? It is unlikely that all long-haul flights (for instance) will be 8000 miles.

14 MAX   M2 + M3 + M7 + N11 + N12 + O7 + O10 + O11 + O12
  + P1 + P2 + P8 + Q4 + Q5 + Q7 + Q8 + R8 + R10 + R11
  + S2 + S3 + S6 + S7 + T1 + T2 + T10 + U8 + U10 + U12
  + V6 + V10 + V12 + W4 + W7 + W8 + W9 + X10 + X11

ST   M2 + M3 + M7 <= 1

   N11 + N12 <= 1

   O7 + O10 + O11 + O12 <= 1

   P1 + P2 + P8 <= 1

   Q4 + Q5 + Q7 + Q8 <= 1

   R8 + R10 + R11 <= 1

   S2 + S3 + S6 + S7 <= 1

   T1 + T2 + T10 <= 1

   U8 + U10 + U12 <= 1

   V6 + V10 + V12 <= 1

   W4 + W7 + W8 + W9 <= 1

   X10 + X11 <= 1

   P1 + T1 <= 1

   M2 + P2 + S2 + T2 <= 1

   M3 + S3 <= 1

   W4 + Q4 <= 1

$$Q5 <= 1$$
$$S6 + V6 <= 1$$
$$M7 + O7 + Q7 + S7 + W7 <= 1$$
$$P8 + Q8 + R8 + U8 + W8 <= 1$$
$$W9 <= 1$$
$$O10 + R10 + T10 + U10 + V10 + X10 <= 1$$
$$N11 + O11 + R11 + X11 <= 1$$
$$N12 + O12 + U12 + V12 <= 1$$

END

This gives 11 different programmes, e.g. 1 pm P, 2 pm T, 3 pm M, 5 pm Q, 6 pm S, 7 pm O, 8 pm, U, 9 pm W, 10 pm V, 11 pm X, 12 pm N; 11 hours

15 (i) MIN $9WA + 5WB + WC + 2WD +$
$4XA + 5XB + XC + 3XD +$
$7YA + 6YB + 2YC + 5YD +$
$4ZA + 7ZB + 3ZC + 2ZD$

ST $WA + WB + WC + WD = 5$
$XA + XB + XC + XD = 5$
$YA + YB + YC + YD = 5$
$ZA + ZB + ZC + ZD = 5$
$WA + XA + YA + ZA = 3$
$WB + XB + YB + ZB = 7$
$WC + XC + YC + ZC = 5$
$WD + XD + YD + ZD = 5$

END

OBJECTIVE FUNCTION VALUE

1)	67.00000	
VARIABLE	VALUE	REDUCED COST
WA	0.000000	5.000000
WB	0.000000	0.000000
WC	5.000000	0.000000
WD	0.000000	0.000000
XA	3.000000	0.000000
XB	2.000000	0.000000
XC	0.000000	0.000000
XD	0.000000	1.000000
YA	0.000000	2.000000
YB	5.000000	0.000000
YC	0.000000	0.000000
YD	0.000000	2.000000
ZA	0.000000	0.000000
ZB	0.000000	2.000000
ZC	0.000000	2.000000
ZD	5.000000	0.000000

(ii) If W is open it supplies 5 lorry loads; if it is closed it supplies 0.

(iii) MIN $9WA + 5WB + WC + 4XA + 5XB + XC$
$+ 7YA + 6YB + 2YC + 4ZA + 7ZB + 3ZC$

ST $WA + WB + WC - 5IW = 0$
$XA + XB + XC - 5IX = 0$
$YA + YB + YC - 5IY = 0$
$ZA + ZB + ZC - 5IZ = 0$
$WA + XA + YA + ZA = 3$
$WB + XB + YB + ZB = 7$
$WC + XC + YC + ZC = 5$
$IW + IX + IY + IZ = 3$

END
INT IW
INT IX
INT IY
INT IZ

Close warehouse Y and supply 5 lorry loads from W to C,
5 lorry loads from X to B, 3 lorry loads from Z to A and
and 2 lorry loads from Z to B.

OBJECTIVE FUNCTION VALUE

1)	56.00000	
VARIABLE	VALUE	REDUCED COST
IW	1.000000	−5.000000
IX	1.000000	−5.000000
IY	0.000000	0.000000
IZ	1.000000	5.000000
WA	0.000000	7.000000
WB	0.000000	0.000000
WC	5.000000	0.000000
XA	0.000000	2.000000
XB	5.000000	0.000000
XC	0.000000	0.000000
YA	0.000000	4.000000
YB	0.000000	0.000000
YC	0.000000	0.000000
ZA	3.000000	0.000000
ZB	2.000000	0.000000
ZC	0.000000	0.000000

16 (i) Introduce edges SX (with capacity 20), SY (20), SZ (15), AT (25) and BT (25).

(ii) 42 million m³ per day

(iii) Potential extra flows (backflows) along path:

SZ 2 (–), ZR 1 (6), RN 10 (0), NQ 10 (0), QM 1 (5), MB 14 (6), BT 9 (–)

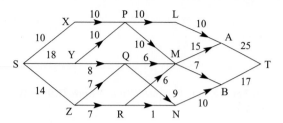

(iv) The flow of 42 equals the capacity of the cut, so it is maximal (by the maximum flow/minimum cut theorem).

(v) MAX   AT + BT

ST   SX − XP = 0

           SY − YP − YQ = 0

           SZ − ZQ − ZR = 0

           XP + YP − PL − PM = 0

           YQ + ZQ − QM − QN = 0

           ZR − RM − RN = 0

           PL − LA = 0

           PM + QM + RM − MA − MB = 0

           QN + RN − NB = 0

           LA + MA − AT = 0

           MB + NB − BT = 0

           SX <= 20

           SY <= 20

           SZ <= 15

           XP <= 18

           YP <= 10

           YQ <= 8

           ZQ <= 7

           ZR <= 7

           PL <= 18

           PM <= 10

           QM <= 6

           QN <= 10

           RM <= 6

           RN <= 10

           LA <= 10

           MA <= 20

           MB <= 20

           NB <= 10

           AT <= 25

           BT <= 25

END

This might give SX = 18, SY = 10 and YP = 2, or MA = 7, MB = 15, AT = 17 and BT = 25.

(vi) Two inequalities have changed:

PL + ML – LA – LM = 0

PM + QM + RM + LM – MA – MB – ML = 0

There are two new inequalities:

LM <= 10

ML <=10

Total flow is now 50, e.g. 8 in LM, 15 in MB and 25 in BT.

17 (i) SABC I DT

(ii)

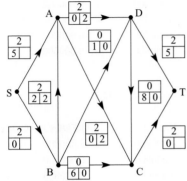

flow	
forward potential	backward potential

(iii) SABDT

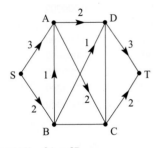

(iv) MAX   SA + SB

ST   SA + BA – AD – AC = 0

SB – BA – BC – BD = 0

BC + AC + DC – CT = 0

AD + BD – DT – DC = 0

SA <= 7

AD <= 2

BC <= 6

CT <= 2

SB <= 2

AC <= 2

DC <= 8

BA <= 4

BD <= 1

DT <= 7

e.g. OBJECTIVE FUNCTION VALUE

1)   5.000000

VARIABLE	VALUE	REDUCED COST
SA	4.000000	0.000000
SB	1.000000	0.000000
BA	0.000000	0.000000
AD	2.000000	0.000000
AC	2.000000	0.000000
BC	0.000000	0.000000
BD	1.000000	0.000000
DC	0.000000	1.000000
CT	2.000000	0.000000
DT	3.000000	0.000000

(v) MIN   7SA + 2SB + 4BA + 2AD + 2AC +
             BD + 6BC + 8DC + 7DT + 2CT

ST   SA + BA − AD − AC = 0
      SB − BA − BC − BD = 0
      BC + AC + DC − CT = 0
      AD + BD − DT − DC = 0
      SA + SB = 1
      DT + CT = 1

END

The shortest route is SBCT, with length 10.

e.g. OBJECTIVE FUNCTION VALUE

1) 10.00000

VARIABLE	VALUE	REDUCED COST
SA	0.000000	1.000000
SB	1.000000	0.000000
BA	0.000000	0.000000
AD	0.000000	5.000000
AC	0.000000	0.000000
BD	0.000000	0.000000
BC	1.000000	0.000000
DC	0.000000	3.000000
DT	0.000000	0.000000
CT	1.000000	0.000000

(vi) The minimum cut is the same as before.

MIN   7SA + 2SB + 4BA + 2AD + 2DA + 2AC
       + BD + 6BC + 8DC + 7DT + 2CT

ST   SA + BA + DA − AD − AC = 0
      SB − BA − BC − BD = 0
      BC + AC + DC − CT = 0
      AD + BD − DA − DT − DC = 0
      SA + SB = 1
      DT + CT = 1

END

The shortest route is SBDACT, with length 9.

OBJECTIVE FUNCTION VALUE

1) 9.000000

VARIABLE	VALUE	REDUCED COST
SA	0.000000	2.000000
SB	1.000000	0.000000
BA	0.000000	1.000000
AD	0.000000	4.000000
DA	1.000000	0.000000
AC	1.000000	0.000000
BD	1.000000	0.000000
BC	0.000000	1.000000
DC	0.000000	4.000000
DT	0.000000	1.000000
CT	1.000000	0.000000

18 (i)

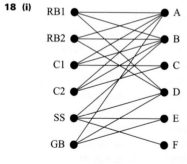

(ii)

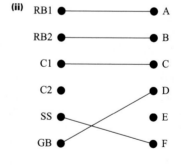

(iii) e.g.

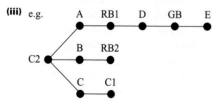

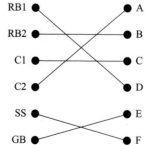

(iv) Putting 1 for RB1, 2 for RB2, etc.:

MAX   A1 + A2 + A3 + A4 + A6 + B1 + B2 + B3 + B4 + B5
      + C3 + C4 + D1 + D2 + D5 + D6 + E5 + E6 + F5

ST   A1 + A2 + A3 + A4 + A6 <= 1
      B1 + B2 + B3 + B4 + B5 <= 1
      C3 + C4 <= 1
      D1 + D2 + D5 + D6 <= 1
      E5 + E6 <= 1
      F5 <= 1
      A1 + B1 + D1 <= 1
      A2 + B2 + D2 <= 1
      A3 + B3 + C3 <= 1
      A4 + B4 + C4 <= 1
      B5 + D5 + E5 + F5 <= 1
      A6 + D6 + E6 <= 1

END

Gives A1, B3, C4, D2, E6, F5 (or A1, B4, C3, D2, E6, F5, or A2, B3, C4, D1, E6, F5, or A2, B4, C3, D1, E6, F5, or A3, B2, C4, D1, E6, F5, or A3, B1, C4, D2, E6, F5, or A4, B2, C3, D1, E6, F5, or A4, B1, C3, D2, E6, F5).

(v) The objective changes to:

MAX   400A1 + 400A2 + 250A3 + 250A4 + 350A6 + 450B1 + 450B2 + 200B3 + 200B4 + 200B5
      + 150C3 + 150C4 + 300D1 + 300D2 + 150D5 + 400D6 + 200E5 + 300E6 + 250F5

Gives £1700 with, for example, A3, B1, C4, D2, E6, F5

19 (i) Using variables A1, A3, etc., where A1 has the value 1 if individual A is allocated to task 1, 0 if not, etc:

MAX   A1 + A3 + B4 + B5 + C1 + C2 + D2 + D4 + D5 + E3 + E4

ST   A1 + A3 = 1
      B4 + B5 = 1
      C1 + C2 = 1
      D2 + D4 + D5 = 1
      E3 + E4 = 1
      A1 + C1 = 1
      C2 + D2 = 1
      A3 + E3 = 1
      B4 + D4 + E4 = 1
      B5 + D5 = 1

(iii) The objective changes to:

MIN   20A1 + 25A3 + 15B4 + 30B5 + 12C1 + 14C2 + 30D2 + 20D4 + 20D5 + 17E3 + 18E4

(ii) e.g. OBJECTIVE FUNCTION VALUE

1) 5.000000

VARIABLE	VALUE	REDUCED COST
A1	0.000000	0.000000
A3	1.000000	0.000000
B4	0.000000	0.000000
B5	1.000000	0.000000
C1	1.000000	0.000000
C2	0.000000	0.000000
D2	1.000000	0.000000
D4	0.000000	0.000000
D5	0.000000	0.000000
E3	0.000000	0.000000
E4	1.000000	0.000000

A does task 3, B does 5, C does 1, D does 2, and E does 4.

(iv) OBJECTIVE FUNCTION VALUE

1) 86.00000

VARIABLE	VALUE	REDUCED COST
A1	1.000000	0.000000
A3	0.000000	0.000000
B4	1.000000	0.000000
B5	0.000000	0.000000
C1	0.000000	0.000000
C2	1.000000	0.000000
D2	0.000000	0.000000
D4	0.000000	0.000000
D5	1.000000	0.000000
E3	1.000000	0.000000
E4	0.000000	0.000000

A does task 1, B does 4, C does 2, D does 5, and E does 3.

Cost = 86

(v) MIN $\quad$ 20A1 + 25A3 + 15B4 + 30B5 + 12C1 + 14C2 +

$\qquad$ 30D2 + 20D4 + 20D5 + 17E3 + 18E4

$\quad$ ST $\quad$ A1 + A3 = 2

$\qquad$ B4 + B5 = 1

$\qquad$ C1 + C2 = 3

$\qquad$ D2 + D4 + D5 = 5

$\qquad$ E3 + E4 = 4

$\qquad$ A1 + C1 = 3

$\qquad$ C2 + D2 = 3

$\qquad$ A3 + E3 = 3

$\qquad$ B4 + D4 + E4 = 3

$\qquad$ B5 + D5 = 3

(vi) OBJECTIVE FUNCTION VALUE

1) $\qquad\qquad$ 274.0000

VARIABLE	VALUE	REDUCED COST
A1	2.000000	0.000000
A3	0.000000	0.000000
B4	1.000000	0.000000
B5	0.000000	0.000000
C1	1.000000	0.000000
C2	2.000000	0.000000
D2	1.000000	0.000000
D4	1.000000	0.000000
D5	3.000000	0.000000
E3	3.000000	0.000000
E4	1.000000	0.000000

A does task 1 twice, B does 4 once, C does 1 once and 2 twice, D does 2 and 4 once each and 5 three times, and E does 3 three times and 4 once. Cost = 274

Chapter 7

❼ (Page 184)

The iteration $x_{n+1} = \dfrac{x_n^2 + 25}{2x_n}$ converges rapidly to 5 from all positive starting values. The iteration $x_{n+1} = \dfrac{x_n^2 - 4}{2x_n}$ does not converge.

Exercise 7A (Page 189)

1 (i) (a) $u_0, u_1 = u_0 + 3, u_2 = u_0 + 6, u_3 = u_0 + 9, u_4 = u_0 + 12$

$\quad$ **(b)** $u_n = u_0 + 3n \qquad\qquad n \geqslant 1$

(ii) (a) $u_4, u_5 = u_4 + 3, u_6 = u_4 + 6, u_7 = u_4 + 9, u_8 = u_4 + 12$

$\quad$ **(b)** $u_n = u_4 + 3(n-4) \qquad n \geqslant 5$

(iii) (a) $u_0, u_1 = 4u_0 + 3, u_2 = 16u_0 + 15, u_3 = 64u_0 + 63, u_4 = 256u_0 + 255$

$\quad$ **(b)** $u_n = 4^n u_0 + 4^n - 1 \qquad n \geqslant 1$

(iv) (a) $u_4, u_5 = 4u_4 + 3, u_6 = 16u_4 + 15, u_7 = 64u_4 + 63, u_8 = 256u_4 + 255$

$\quad$ **(b)** $u_n = 4^{n-4} u_0 + 4^{n-4} - 1 \qquad n \geqslant 5$

2 £1490.29

3 $u_{n+1} = u_n + (10 \times 2^n)$ $\qquad u_0 = 7$

$\quad u_n = (10 \times 2^n) - 3$ $\qquad\quad n \geqslant 0$

4 169 seconds

5 (i) L(3, 2, 1) M(−) R(−) → L(3, 2) M(−) R(1) → L(3) M(2) R(1) → L(3) M(2, 1) R(−) → L(−)M(2, 1) R(3)

$\quad$ → L(1) M(2) R(3) → L(1) M(−) R(3, 2) → L(−) M(−) R(3, 2, 1)

$\quad$ **(ii)** $2^{64} - 1 \approx 18\,400\,000\,000\,000\,000\,000$ days

6 (i) $u_{n+1} = 400 + 0.8u_n$

$\quad$ **(ii)** $u_n = 2000(1 - (0.8)^n)$ or $2000 - 1600(0.8)^{n-1}$

$\quad$ **(iii) (a)** $u_{12} = £1862.56$

$\qquad\quad$ **(b)** £2000

Exercise 7B (Page 195)

1 (i) $u_n = A2^n + Bn2^n$

$\quad$ **(ii)** $u_n = A3^n + Bn3^n$

2 $u_n = 3^{n-2} + 2n3^{n-2}$

3 $u_n = \dfrac{(6 \times 4^n) + (-3)^n}{7}$

4 $u_n = \dfrac{1}{\sqrt{5}} \left[\left(\dfrac{1+\sqrt{5}}{2} \right)^{n+1} - \left(\dfrac{1-\sqrt{5}}{2} \right)^{n+1} \right]$

5 10 946 ways (the 20th Fibonnaci number)

$\quad$ (You can form a path of length n by adding a 'sideways-on' slab to a path of length $n-1$, or by adding two 'end-on' slabs to a path of length $n-2$.)

6 $u_n = (-9 \times 2^n) + (3 \times 3^n) + 0.5n^2 + 3.5n + 8$

Exercise 7C (Page 196)

1 (i) $u_n = \left(c + \dfrac{b}{a-1} \right) a^n - \dfrac{b}{a-1}$ or $a^n c + b \dfrac{(a^n - 1)}{(a-1)}$

$\quad$ **(ii)**

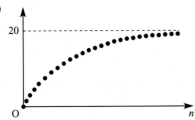

2 (i) (a) $u_n = -\frac{2}{7}(-4)^n + \frac{2}{7}3^n$

$\qquad$ **(b)** $u_2 = -2;\ u_3 = 26;\ u_4 = -50$

$\qquad$ **(c)** $-\frac{2}{7}(-4)^4 + \frac{2}{7}3^4 = -73\frac{1}{7} + 23\frac{1}{7} = -50$

$\quad$ **(ii) (a)** $12 \times (-2) - (-0.2) + 2 = -0.2$

$\qquad$ **(b)** $v_n = \frac{8}{35}(-4)^n + \frac{15}{35}3^n - 0.2$

3 (i) $v_k = 0.99v_{k-1} + 0.1$

$\quad$ Divide through by 10 to get concentrations, and use $c_k = \dfrac{v_k}{10}$.

$\quad$ **(ii)** $c_k = 1 + 0.99^k(c_0 - 1)$

$\qquad$ $c_4 \approx 0.294$. It is ten weeks before the concentration reaches 0.5.

4 (i) 2.5%

   (ii) 25 ml

   (iii) 475 ml.

      4.75% or 0.0475

   (iv) $u_{n+1} = 0.9u_n + 250$

   (v) $c_{n+1} = 0.9c_n + 2.5$ (or $+ 0.025$)

   (vi) 25% or 0.25

5 (i) Add on interest and take off payment. Interest is added by multiplying by 1, to represent the outstanding balance, plus the monthly interest rate, expressed as a decimal.

   (ii) $u_n = 30\,000 \times 1.0095^n - \left(\dfrac{1.0095^n - 1}{0.0095} \right) X$

   (iii) £302.75

6 (i) **(a)** £2520

      (b) £3045.20

   (ii) **(a)** $0.25 \times £2520$

      (b) £761.30

   (iii) 1.01 to add on interest

      0.25 for withdrawal

      £1000 income

   (iv) £$\frac{1000}{0.24}$ = £4166.67

      £1041.67

   (v) £$\frac{1000}{0.49}$ = £2040.82

      £1020.41

7 (i) Two sides (at two matches per square) + a corner to be added on.

      $u_{n+1} = u_n + 4n + 4$

   (ii) $u_n = 2n(n + 1)$ $u_1 = 4$

   (iii) 84

   (iv) $s_{n+1} = s_n + 2(n + 1)(n + 2)$ $s_1 = 4$

8 (i) $u_{n+1} = (1 - b - d)u_n - c$

   (ii) $u_n = p^{n-1}u_1 + q\,\dfrac{(p^{n-1} - 1)}{(p - 1)}$

   (iii) 2000

   (iv) Unlimited population growth; a reducing rate of population growth as resources are depleted.

9 (i) **(a)** 11, 16, 22

      (b) $u_{n+1} = u_n + n + 1$

           Number of regions created by $(n + 1)$th line $= 1 + 1$ more each time an existing line is crossed.

      (c) $u_n = 0.5n^2 + 0.5n + 1$

   (ii) $u_{n+1} = u_n + n$

      Number of new intersections $\leqslant$ number of existing lines.

      $u_n = \dfrac{n(n - 1)}{2}$

10 (i) $Q_{n+1} = Q_n +$ number joining $-$ number leaving

              $= Q_n + (900 - 1.75Q_n) - 200$

              $= 700 - 0.75Q_n$

   (ii) $Q_n = 400 + 100(-0.75)^n$

   (iii) 400

   (iv) The numbers would be smaller and so there would be greater variability and less predictability.

11 (i) $u_3 = 12$

(ii) u_n = existing number

u_{n-1} = number at least one year old, and therefore number of new branches

(iii) (a) $u_n = \dfrac{4}{\sqrt{5}}\left[\left(\dfrac{1+\sqrt{5}}{2}\right)^{n+1} - \left(\dfrac{1-\sqrt{5}}{2}\right)^{n+1}\right]$

(b) $u_2 = \dfrac{4}{\sqrt{5}}\left[\left(\dfrac{1+\sqrt{5}}{2}\right)^{3} - \left(\dfrac{1-\sqrt{5}}{2}\right)^{3}\right]$

$= \dfrac{1}{2\sqrt{5}}\left[1+3\sqrt{5}+15+5\sqrt{5} - \left(1-3\sqrt{5}+15-5\sqrt{5}\right)\right]$

$= 8$

(iv) (a) $u_5 = 32$; $u_4 = 20$.

(b) 9

(c) 239, 29 411

(d) 23 are old and 11 are new.

Removing 14 leads to 127; removing 15 leads to 119

12 (i) $u_{n+1} = 0.8u_n + 20\,000$ $u_0 = 250\,000$

(ii) $u_5 = 149\,152$

$u_n = 150\,000 \times 0.8^n + 100\,000$

(iii) 16 years

(iv) 100 000 tonnes

(v) Recurrence relation as before, but now $u_3 = 56\,800$, not 176 800.

New $u_5 = 72\,352$

New $u_n = 100\,000 - 84\,375 \times 0.8^n$ $n \geqslant 3$

Long run capacity requirement is still 100 000 tonnes.

Quantity of stored seed is now *increasing* year by year.

13 (i) New length = length + growth

(ii) $L_{n+1} = L_n + 0.9(L_n - L_{n-1}) = 1.8L_n - 0.8L_{n-1}$

(iii) $L_n = 6 - 5 \times 0.8^n$

(iv) approximately every 4 days

(v) 50 times at 3 cm gives 150 cm.

(vi) Cutting at $L = 4.7$ gives about 6 days between cuts.

This gives 33 cuts per year, so about 122 or 123 cm.

14 (i) $w_1 = 1$ (for Indiana's final day) + 1 (for the remaining porter's walk back to C_1) = 2

(ii) $w_{k+1} = w_k$ (required at C_k) + $k + 1$ (to get it there, k porters + Indiana)

  + 1 (for the returning porter to get back to C_{k+2})

  + k (for the eventually returning porters to get back to C_{k+2})

(iii) $w_n = n(n+1)$

(iv) $w_{19} = 380$

So 380 + 20 = 400 cans are needed

(v) $y_0 = 1, y_1 = 4$

(vi) $y_{k+1} = y_k + 2k + 3$

(vii) $y_n = (n+1)^2$

$y_{19} = 400$, so they need to set out with 420 cans (or 400 from part (iv) + 20 for Indiana's return)

15 (i) (a) $y_n = 6.4x_{n-1} - 36$; $x_{n+1} = 2.56x_{n-1} - 23.4$

(b) $x_1 = 0.4 \times 60 - 9 = 15$

The auxiliary equation has solutions ±1.6, the particular solution is 15, and the initial conditions of 17 and 15 give both constants equal to 1.

$x_n = (1.6)^n + (-1.6)^n + 15$

(c) $x_2 = 20.12, x_3 = 15, x_4 = 28.1072, x_5 = 15, x_6 = 48.554432, x_7 = 15, x_8 = 100.899\,345\,9, x_9 = 15, x_{10} = 234.902\,325\,6$

(ii) (a)

Month	Number of breeding pairs	Number of eggs
0	17	60
1	15	72.8
2	20.12	60
3	15	92.768
4	28.1072	60
5	15	143.886 08
6	48.554 432	60
7	15	274.748 364 8
8	100.899 345 9	60
9	15	609.755 813 9

(b)

Month	Number of breeding pairs	Number of eggs	Month	Number of breeding pairs	Number of eggs
0	17	60	0	14	60
1	15	72	1	15	53
2	19	60	2	12	60
3	15	85	3	15	40
4	25	60	4	7	60
5	15	124	5	15	8
6	40	60	6	0	60
7	15	220	7	15	0
8	79	60	8	0	60
9	15	469	9	15	0

Initial numbers of 15 breeding pairs and 60 eggs give an equilibrium.

16 (i) $4.8 - 0.5 = 4.3$

(ii) 4.5, 5.2, 5.7, 5.5, 4.8, 4.3; this pattern is repeated in subsequent years.

(iii), (iv), (v) The formula for an entry in column C is '=MAX(C2+1.25*(5–C1),0)'.

A	B	C	D	E	A	B	C	D	E
0	4.8	4.80000	4.80000	4.80000	20	4.5	7.68636	4.97697	4.89666
1	4.3	4.30000	4.30000	4.30000	21	5.2	11.58512	4.95783	4.90699
2	4.5	4.55000	4.45000	4.31800	22	5.7	8.22718	4.97510	4.91629
3	5.2	5.42500	4.97500	4.38100	23	5.5	0.00000	5.00673	4.92467
4	5.7	5.98750	5.38750	4.44238	24	4.8	0.00000	5.02540	4.93220
5	5.5	5.45625	5.40625	4.49809	25	4.3	6.25000	5.02036	4.93898
6	4.8	4.22188	5.11563	4.54828	26	4.5	12.50000	5.00130	4.94508
7	4.3	3.65156	4.81094	4.59345	27	5.2	10.93750	4.98604	4.95057
8	4.5	4.62422	4.72422	4.63410	28	5.7	1.56250	4.98506	4.95552
9	5.2	6.30977	4.86602	4.67069	29	5.5	0.00000	4.99553	4.95996
10	5.7	6.77949	5.07285	4.70362	30	4.8	4.29688	5.00674	4.96397
11	5.5	5.14229	5.17334	4.73326	31	4.3	10.54688	5.01009	4.96757
12	4.8	2.91792	5.11870	4.75993	32	4.5	11.42578	5.00504	4.97081
13	4.3	2.74006	4.98870	4.78394	33	5.2	4.49219	4.99747	4.97373
14	4.5	5.34266	4.89967	4.80555	34	5.7	0.00000	4.99369	4.97636
15	5.2	8.16758	4.90815	4.82499	35	5.5	0.63477	4.99559	4.97872
16	5.7	7.73925	4.98340	4.84249	36	4.8	6.88477	5.00032	4.98085
17	5.5	3.77977	5.05228	4.85824	37	4.3	12.34131	5.00363	4.98277
18	4.8	0.35571	5.06474	4.87242	38	4.5	9.98535	5.00339	4.98449
19	4.3	1.88099	5.02552	4.88518	39	5.2	0.80872	5.00067	4.98604
					40	5.7	0.00000	4.99813	4.98744

(vi) The equation is $u_{n+2} - u_{n+1} + 0.09u_n = 0.45$.

The auxiliary equation is $m^2 - m + 0.09 = 0$, giving $m = 0.9$ or 0.1.

The particular solution is 5, so $u_n = A0.9^n + B0.1^n + 5$.

The initial conditions give $A + B = -0.2$ and $0.9A + 0.1B = -0.7$.

Hence $A = -\frac{68}{80}$ (−0.85) and $B = \frac{52}{80}$ (0.65).

(vii) The results are shown in column E above: this policy gives non-oscillatory convergence to 5 (slower than in part (v)).

17 (i) $0.75 =$ remainder after excretion

$200 - x_n =$ top-up dose

$150 = 0.75 \times 200$ and $112.5 = 0.75 \times 150$

(ii), (iii), (iv) Results shown in columns A, B and C respectively.

The therapy described in part (iii) does not help: the amount just fluctuates at a higher level.

Using the therapy described in part (iv) the amount of drug converges to a level of 133.33 ... .

A	B	C	D	E	F
1	150.0	150.0	150.0	150.0	190.00
2	112.5	112.5	112.5	162.5	182.50
3	134.4	184.4	109.4	159.4	176.88
4	188.3	275.8	125.8	160.2	172.66
5	206.8	272.5	139.6	160.0	169.49
6	166.8	204.3	141.8	160.0	167.12
7	125.1	153.3	136.6	160.0	165.34
8	127.0	160.6	131.5	160.0	164.00
9	170.1	217.2	130.3	160.0	163.00
10	200.6	252.3	132.0	160.0	162.25
11	180.3	222.0	133.8	160.0	161.69
12	135.2	166.5	134.4	160.0	161.27
13	121.1	152.9	133.9	160.0	160.95
14	155.6	198.1	133.2	160.0	160.71
15	195.6	245.7	133.0	160.0	160.53
16	191.1	236.2	133.1	160.0	160.40
17	147.7	181.4	133.4	160.0	160.30
18	119.7	149.9	133.5	160.0	160.23
19	142.0	181.0	133.4	160.0	160.17
20	186.8	235.9	133.3	160.0	160.13
21	198.1	245.9	133.3	160.0	160.10

(v) $x_n = 160\left(1 - \left(-\frac{1}{4}\right)^{n+1}\right)$

The results are shown in column E above: with this therapy the drug level quickly converges to a level of 160.

(vi) The results are shown in column F above: with this therapy the level also converges to 160 – less quickly, but no tests are needed.

18 (i) (a) $24 - 0.2 \times 24 - 0.15 \times 6 = 18.3$

(b) $u_{n+2} = 0.8u_{n+1} + 0.15(u_{n+1} - u_n)$

(c) The auxiliary equation is $x^2 - 0.95x + 0.15 = 0$, with solution $x = 0.2$ or $x = 0.75$.

The general solution is $u_n = A(0.2)^n + B(0.75)^n$.

$A + B = 30$ and $0.2A + 0.75B = 24$, hence $A = -\frac{30}{11}$ (−2.7272...) and $B = \frac{360}{11}$ (32.7272...).

(d) The bunny approaches the burrow.

(ii) (a) $u_{n+2} = 0.75u_{n+1} + 0.25(u_{n+1} - u_n)$

(b) The auxiliary equation is $x^2 - x + 0.25 = 0$, with solution $x = 0.5$ (repeated).

The general solution is $u_n = A(0.5)^n + Bn(0.5)^n$.

$A = 30$ and $B = 15$.

(c) The bunny approaches the burrow more quickly.

(iii) (a) $u_{n+2} - u_{n+1} + 0.5u_n = 0$

(b) The bunny hops in and out of the burrow alternately to the left and right, and by decreasing amounts.

30	30
15	15
=0.5* A2+0.5*(A2–A1)	0
=0.5*A3+0.5*(A3–A2)	–7.5
etc.	–7.5
	–3.75
	0
	1.875
	1.875
	0.9375
	0
	–0.46875
	–0.46875
	–0.23438
	0
	0.117188
	0.117188
	0.058594
	0
	–0.0293

19 (i) 10-second temperature values are:

$21 \to 23 \to 25 \to 27 \to 29 \to 31 \to 33 \to 35 \to 37 \to 39 \to 37 \to 39 \to \ldots$

The temperature increases for the first 90 seconds and then oscillates above and below the required temperature.

(ii) 5-second temperature values are:

$21 \to 21 \to 23 \to 25 \to 27 \to 29 \to 31 \to 33 \to 35 \to 37 \to 39 \to 41 \to 39 \to 37 \to 35 \to 37 \to \ldots$

(iii) u_n = temperature at time $5n$

$\left(\frac{1}{2}\right)^{n-1} = \left(\frac{1}{2}\right)^n \times 2$

x_n indicates whether the water is too hot or too cold – whether to adjust down or up.

Adjustment is defined 10 seconds before (at step $n-2$) but affects current temperature as defined 5 seconds before (at step $n-1$).

(iv)

Time	Temperature	x_n
0	21	–1
1	21	–1
2	23	–1
3	24	–1
4	24.5	–1
5	24.75	–1
6	24.875	–1
7	24.9375	–1
8	24.96875	–1
9	24.98438	–1
10	24.99219	–1
11	24.99609	–1
12	24.99805	–1

The temperature converges to 25°C.

(v) Initial adjustment is 9°C. Thereafter the adjustment is reduced by a factor of 2 at each iteration.

(vi)

Time	Temperature	x_n
0	21	−1
1	21	−1
2	30	−1
3	34.5	−1
4	36.75	−1
5	37.875	−1
6	38.4375	1
7	38.71875	1
8	38.57813	1
9	38.50781	1
10	38.47266	1
11	38.45508	1
12	38.44629	1

The temperature converges to $38\frac{7}{16}°$C.

(vii)

Time	Temperature
0	21
1	21
2	29.5
3	38
4	42.25
5	42.25
6	40.125
7	38
8	36.9375
9	36.9375
10	37.46875
11	38
12	38.265625

There is oscillatory convergence to 38°C.

Chapter 8

❼ (Page 209)

The contestant *should* switch his choice. He doubles his chance of winning by doing so. If his first choice is correct, which happens with probability $\frac{1}{3}$, then the host has a choice of which door to throw open. The contestant loses by switching. If the contestant's first choice is incorrect, which happens with probability $\frac{2}{3}$, then the host throws open the remaining incorrect door, leaving the prize behind the unselected door. The contestant wins by switching.
So switching wins two times out of three and loses once out of three.
The argument is irrefutable and the simulation is convincing.

(Page 213)

The queue length builds up steadily, if slowly.

📖 **(Page 216)**

The graph shows that you only have to be slightly worse than your opponent to have very little chance of winning.

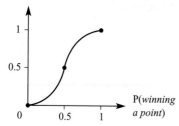

Exercise 8A (Page 219)

1 e.g. (Column headings in row 1, so calculations start in row 2.)

Column headings	Formulae
A: Customer	As figure 8.5
B: Arrival time	As figure 8.5
C: Server	'1' in row 2. Then '=IF(G2<F2,2,1)'
D: Start of service	'=B2' in row 2. Then '=IF(C3=1,MAX(B3,F2),MAX(B3,G2))'
E: Service time	As figure 8.5
F: End server 1	'=D2+E2' in row 2. Then '=IF(C3=1,D3+E3,F2)'
G: End server 2	'0' in row 2. Then '=IF(C3=2,D3+E3,G2)'
H: Q time	'= D2–B2' in row 2, 'D3–B3' in row 3, etc.
I: Mean Q time	Empty in row 2, then '=SUM(H$2:H3)/D3'
J: Server 1 idle	Empty in row 2, then '=IF(C3=1,D3-F2,'–')'
K: Server 1 utilisation	Empty in row 2, then '=1-SUM(J$3:J3)/F3
L: Server 2 idle	Empty in row 2, then '=IF(C3=2,D3-G2,'–')'
M: Server 2 utilisation	Empty in row 2, then '=1-SUM(L$3:L3)/G3'

2 e.g. (Column headings in row 1, so calculations start in row 2.)

Column headings	Formulae
Columns A, B and C	As figure 8.6
D: Points A	Row 4 changes to '=IF(L3=1,0,IF(C4<B$1,1+D3,D3))'
E: Points B	Row 4 changes to '=IF(L3=1,0,IF(C4>B$1,1+E3,E3))'
F: Games A	Row 4 changes to '=IF(M3=1,0,IF(AND(D4>(E4+1),D4>3),1+F3,F3))'
G: Games B	Row 4 changes to '=IF(M3=1,0,IF(AND(E4>(D4+1),E4>3),1+G3,G3))'
H: Sets A	Row 4 changes to '=IF(N3=1,0,IF(AND(F4>(G4+1),F4>5),1+H3,H3))'
I: Sets B	Row 4 changes to '=IF(N3=1,0,IF(AND(G4>(F4+1),G4>5),1+I3,I3))'
J: Matches A	'0' in row 2, then '=IF(H3=3,J2+1,J2)'
K: Matches B	'0' in row 2, then '=IF(I3=3,K2+1,K2)'
L: (game flag)	Empty in row 2, then '=IF(OR(F3>F2,G3>G2),1,0)'
M: (set flag)	Empty in row 2, then '=IF(OR(H3>H2,I3>I2),1,0)'
N: (match flag)	Empty in row 2, then '=IF(OR(J3>J2,K3>K2),1,0)'

3 A flag column can be set up to indicate who is serving. Probabilities can then be placed in column B using an IF formula, so that the values depend on the value of the serving flag. You will then need to delete the $ from B$1 in columns D and E *before* copying down. The referencing will then be to the matching row of column B.

4 Convergence to π is slow. Many repetitions are needed to produce a reliably accurate estimate.

5 Expect to find about 37 bottles with no pebbles, 37 with one pebble, 18 with two pebbles and about 6 with three pebbles

❷ (Page 220)

The fact that the total of the probabilities is 1 tells you that all possibilities are covered, so the only values X can take are 1, 2, 3 and 4.

❷ (Page 221)

Imagine running a simulation of a queue and observing the length of the queue after, say, 5 seconds and again after 6 seconds. It is clear that the length of the queue after 6 seconds will not be independent of the length of the queue after 5 seconds. For observations to be independent, there must be a reasonably long time interval, for example one minute, between them.

❷ (Page 222)

1 They have been reduced.

2 $0.656 = \dfrac{0.927}{\sqrt{2}}$

Exercise 8B (Page 223)

1 (i) 21 (rounding up to be safe)
  (ii) 82
  (iii) 2026

2 (i) **(a)** 33
     (b) 129
     (c) 3212
  (ii) 2.1
  (iii) **(a)** 0.355
       (b) 0.180
       (c) 0.036

Exercise 8C (Page 225)

1 (i) **(a)** e.g.
                    00 to 55 A wins
                    56 to 99 B wins
     First eight points are A, A, B, A, B, B, A, A.
     Score goes 1–0, 2–0, 2–1, 3–1, 3–2, 3–3, 4–3, 5–3.
     A wins 5–3.
    (b) Repeat many times to estimate the probability of A winning.
  (ii) See figure 8.6.

(iii)

	A	B	C	D
1	p=	0.28		Points A
2		0.22		0
3			=RAND()	=IF(C3<B$1,1+D2,D2)
4			=RAND()	=IF(J3=1,0,IF(C4<B$1,1+D3,D3))

	E	F
1	Points B	Points C
2	0	0
3	=IF(AND(C3>B$1,C3<B$1+B$2),1+E2,E2)	=IF(C3>B$1+B$2,1+F2,F2)
4	=IF(J3=1,0,IF(AND(C4>B$1,C4<B$1+B$2),1+E3,E3))	=IF(J3=1,0,IF(C4>B$1+B$2,1+F3,F3))

	G	H
1	Games A	Games B
2	0	0
3	=IF(AND(D3>(E3+1),D3>(F3+1),D3>5),1+G2, G2)	=IF(AND(E3>(D3+1),E3>(F3+1),E3>5),1+H2,H2)
4	=IF(AND(D4>(E4+1),D4>(F4+1),D4>5),1+G3, G3)	=IF(AND(E4>(D4+1),E4>(F4+1),E4>5),1+H3,H3)

	I	J
1	Games C	Flag
2	0	0
3	=IF(AND(F3>(D3+1),F3>(E3+1),F3>5),1+I2,I2)	=IF(OR(G3>G2,H3>H2,I3>I2),1,0)
4	=IF(AND(F4>(D4+1),F4>(E4+1),F4>5),1+I3,I3)	=IF(OR(G4>G3,H4>H3,I4>I3),1,0)

2 (i) e.g.

	A	B	C	D	E	F
1	=IF(RAND()<0.15,0,1)	=IF(RAND()<0.15,0,1)	etc.	etc.	etc.	=SUM(A1:E1)
2	=IF(RAND()<0.15,0,1)	=IF(RAND()<0.15,0,1)	etc.	etc.	etc.	=SUM(A2:E2)
3	=IF(RAND()<0.15,0,1)	=IF(RAND()<0.15,0,1)	etc.	etc.	etc.	=SUM(A3:E3)
4	=IF(RAND()<0.15,0,1)	=IF(RAND()<0.15,0,1)	etc.	etc.	etc.	=SUM(A4:E4)
etc.	etc.	etc.	etc.	etc.	etc.	etc.

(ii) (a) Such a simulation experiment yielded 94 out of 200 companies with all spaces occupied, giving a probability of 0.47.

(b) The same experiment had an average of 4.28 spaces occupied per company.

(iii) $0.85^5 \approx 0.4437$

3 (i), (ii) The service time and inter-arrival time distributions can be input on sheet 2 of the spreadsheet in figure 8.5.

(iii) One simulation experiment gave $(4 + 5 + 6 + 7 + 5 + 6 + 6 + 4 + 4 + 3 + 0 + 0 + 1 + 0 + 0 + 0) \div (56 - 12) = 1.16$. This takes into account a settling-down period, when queues are unrepresentatively small.

(iv) One simulation experiment gave times 42 to 44, 49 to 50, 53 to 54 and 55 to 56. Utilisation $= \frac{41}{46} \times 100 = 89\%$.

(v) Ten repetitions of the simulation and of the computation in part (iii), gave mean queue lengths with a mean of 1.43 and a standard deviation of 1.13, indicating that the mean is about $1.43 \pm \frac{2.26}{\sqrt{10}} \approx 1.43 \pm 0.71$.

(vi) Ten repetitions gave a mean of 95.9% with a standard deviation of 4.54%. So the mean percentage utilisation is about $95.9 \pm \frac{9.1}{\sqrt{10}} \approx 95.9 \pm 2.9$.

4 (i) **(a)** 1.5–2.5, 5.5–7.5, 10.5–12.5, 15.5–17.5, 20.5–22.5, 25.5–27.5, 30.5–32.5, 35.5–37.5, 40.5–42.5, 45.5–47.5, 50.5–51.5.

(b) Regard each 5-digit random number as 0.*****. Multiply by 50 and add 1.5 to simulate the position of the centre. One simulation experiment gave the following x values: 34.12, 41.54, 42.79, 22.04, 45.14, 4.78, 34.05, 38.47, 2.34, 19.71.

(c) The results for the experiment above would be: L, W, L, W, L, L, L, L, W, L.

(ii) **(b)** Length of valid x interval $= 50$
Length of losing x interval $= 10 \times 3 = 30$
Probability of winning $= 0.4$

(iii) 4.5

5 (i) e.g. Use 3 single-digit numbers. If first number is 0, 1, 2, 3 or 4 then go left at separator 1; otherwise go right. Similarly for next two digits at levels 2 and 3.
One experiment gave the following final destinations (the routes of the first five balls are also shown):
135C; 124B; 124A; 136C; 124B; A; B; A; C; B; B; B; C; C; B; C; C; B; D; B; B; B; C; A

(ii) **(a)** For the experiment above, achieved mean $= (4 \times 5 + 11 \times 10 + 8 \times 10 + 1 \times 5) \div 24 = 8.96$

(b) Expected mean $= (3 \times 5 + 9 \times 10 + 9 \times 10 + 3 \times 5) \div 24 = 8.75$

6 (i) **(a)** 1, 0, 3, 1, 0, 2, 0, 2, 1, 1

(b) 0.7

(c) e.g.　00 to 13 nut centre
14 to 97 not a nut centre
98 and 99 ignore

(ii) **(a)** e.g.　=IF(RAND( )<1/7, 1,0)

(c) The formula '=IF(L1<2, 1,0)', or equivalent, can be used, where L1 contains the total for carton 1. A full 50-row spreadsheet gave 27 boxes containing fewer than two nut centres and so a probability of 0.54.

(iii) **(a)** $(1 + 9 \times 0.15)(0.85)^9 = 0.5443$
$(1 + \frac{9}{7})(\frac{6}{7})^9 = 0.5708$

(b) The first result was a long way off. The low level of accuracy being due to the small number of repetitions. The second result, with fifty repetitions, was much closer.

(c) Use more repetitions. 1000 repetitions gave 569 boxes, i.e. a probability of 0.569.

7 (i) $6 \times 30 = 180$
$5 \times 30 - 1 \times 10 = 140$

(ii) A: =RAND()
B: =LOOKUP(A1,Sheet2!A$1:B$6)　(or by nested 'IF' statements)

Lookup table		
	0	5
	0.1	6
	0.3	7
	0.6	8
	0.9	9
	0.95	10

C: =IF(B1>=6,6*0.3,B1*0.3–(6–B1)*0.1)

(iii) e.g.

0.967224	10	1.8	2.1	2.4	2.7	3
0.794852	8	1.8	2.1	2.4	2.3	2.2
0.45679	7	1.8	2.1	2	1.9	1.8
0.34215	7	1.8	2.1	2	1.9	1.8
0.465551	7	1.8	2.1	2	1.9	1.8
0.387285	7	1.8	2.1	2	1.9	1.8
0.733934	8	1.8	2.1	2.4	2.3	2.2
0.179176	6	1.8	1.7	1.6	1.5	1.4
0.896466	8	1.8	2.1	2.4	2.3	2.2
0.400182	7	1.8	2.1	2	1.9	1.8
0.400022	7	1.8	2.1	2	1.9	1.8
0.828997	8	1.8	2.1	2.4	2.3	2.2
0.788919	8	1.8	2.1	2.4	2.3	2.2
0.185122	6	1.8	1.7	1.6	1.5	1.4
0.298555	6	1.8	1.7	1.6	1.5	1.4
0.322082	7	1.8	2.1	2	1.9	1.8
0.378762	7	1.8	2.1	2	1.9	1.8
0.03436	5	1.4	1.3	1.2	1.1	1
0.660212	8	1.8	2.1	2.4	2.3	2.2
0.149588	6	1.8	1.7	1.6	1.5	1.4

(iv) e.g. £1.76 for 6 copies, £1.94 for 7, £2 for 8, £1.94 for 9, £1.86 for 10.

(v) $\sigma \approx 0.4$. $\dfrac{\sigma}{\sqrt{100}} \approx 0.04$. So accuracy about ± 0.08.

(vi) Require n such that $\dfrac{0.4}{\sqrt{n}} \approx \dfrac{0.02}{2}$, i.e. about 1600 days.

8 **(i)** A: =RAND()

B: =LOOKUP(A1,Sheet2!A$1:B$6), or similar

Lookup table	0	1
	0.5	2
	0.75	3
	0.9	4
	0.95	5
	0.98	10

C: =B2+C1 (after row 1)

(ii) D: =IF(D1<0.4,1,IF(D1<0.7,2,IF(D1<0.9,3,4))), or similar, or =LOOKUP

(iii) =IF(C1<=120,E1,0)

=SUM(F1:F100)

One such simulation gave the following spreadsheet:

A	B	C	D	E	F	G	H
0.991721	10	10	0.905967	4	4		102
0.709742	2	12	0.583360	2	2		
0.392983	1	13	0.695456	2	2		
0.282399	1	14	0.655104	2	2		
0.985256	10	24	0.409523	2	2		
etc.							

(iv) Ten simulations gave totals of 102, 91, 128, 124, 114, 124, 115, 136, 126 and 122; mean = 118.2.

(v) The experiment above gave $s = 13.2732$ and $\dfrac{s}{\sqrt{10}} \approx 4.20$. You require the standard error to be about 2.5, so n needs to be about 30. (These results are computed from the illustrated experiment. In fact the expected values for the mean and standard deviation are about 114 and 17 respectively. This would give a standard error of about 5.4 with $n = 10$, and would require n to be approaching 50 to bring the standard error down to 2.5.)

9 (i) A, B: =RAND()

  C: =A1^2+B1^2

  D: =IF(C1<1,1,0)

(ii) =0.004*SUM(D1:D1000)

  e.g. 3.084

(iii) All points (A, B) lie within the unit square. Those points for which $A^2 + B^2 < 1$ lie within a quadrant of the unit circle. So $\dfrac{\text{sum}}{1000} \approx \dfrac{\pi}{4}$, giving $\pi \approx 0.004 \times \text{sum}$.

(iv) e.g. 3.084, 3.152, 3.096, 3.240, 3.124, 3.116, 3.128, 3.120, 3.188, 3.164, 3.144, 3.140; mean = 3.141 33, σ = 0.042 07.

(v) e.g. You require n such that $\dfrac{0.042}{\sqrt{n}} = \dfrac{0.0005}{2}$, i.e. $n \approx 30\,000$.

(vi) e.g. 3.216, 3.048, 3.186, 3.078, 3.138, 3.168, 3.246, 3.096, 3.168, 3.114, 3.276, 3.282; mean = 3.168, σ = 0.076 79. Comparing standard deviations, and hence standard errors, the new method seems to be worse.

10 (i) (a) First gap more than 8 seconds is at time 30 seconds.

  (b) You could use LOOKUP or nested IF statements.

  (c) e.g. 38, 14, 89, 74, 32, 116, 32, 26, 47 (all are 8 plus a multiple of 3)

  (d) e.g. 52.0 seconds

  (e) e.g. σ = 33.7, so $n = 4\sigma^2 \approx 5000$

(ii) (a) First gap more than 4 seconds is at time 3 seconds, so get to island at time 7 seconds. Then first gap in eastbound traffic is at time 9 seconds, so get to other side at time 13 seconds.

  (c) e.g. 13, 13, 16, 8, 13, 22, 8, 16, 13

  (d) e.g. 13.6 seconds

  (e) e.g. σ = 4.3, n = 9, $\dfrac{2\sigma}{\sqrt{n}} \approx 2.9$, so the average time will lie between about 11 seconds and about 17 seconds.

(iii) Time to cross is very variable without the island, and is certainly much longer on average than it takes with the island.

11 (i) e.g.

Time t (hours)	RAND()	$0.02x(t)y(t)$	$x(t)$	$y(t)$
0			1	5
1	0.85819169	0.1	1	5
2	0.51188945	0.1	1	5
3	0.47759187	0.1	1	5
4	0.6410667	0.1	1	5
...				
49	0.05446689	0.1	6	0
50	0.13065527	0	6	0

In this simulation, all members of the community were infected after 49 hours.

(ii) e.g. 49, 48, 30, 41, 31, 23, 14, 58, 28, 51; mean = 37.3, σ = 14.173 92. So $n \approx \left(\dfrac{2 \times 14.174}{3.73}\right)^2 \approx 60$.

(iii) e.g.

Time t	RAND()	$0.02x(t)y(t)$	Infect?	RAND()	$0.1x(t)$	Recover?	$x(t)$	$y(t)$	$z(t)$
0							1	5	0
1	0.0883	0.1	1	0.1451	0.1	0	2	4	0
2	0.1668	0.16	0	0.0442	0.2	1	1	4	1
3	0.3686	0.08	0	0.7751	0.1	0	1	4	1
4	0.8634	0.08	0	0.6212	0.1	0	1	4	1
...									
18	0.64	0.04	0	0.0107	0.1	1	0	2	4
19	0.3984	0	0	0.8494	0	0	0	2	4

(iv) e.g.

t	x	y	z
18	0	2	4
15	0	3	3
44	0	2	4
27	0	1	5
8	0	5	1
23	0	5	1
12	0	3	3
2	0	5	1
10	0	5	1

It is highly likely that some will survive uninfected (but *not* certain).

Index